D1742000

Corrosion in civil engineering

Proceedings of the conference held in London,
21-22 February, 1979

Institution of Civil Engineers, London, 1979

Conference sponsored by the Institution of Corrosion Science and Technology and the Institution of Civil Engineers

Organizing Committee: J. T. Calvert (Chairman), R. R. Bishop, P. Pullar-Strecker and D. Lewis

Published by the Institution of Civil Engineers, P.O. Box 101, 26-34 Old Street, London EC1P 1JH

Production Editor: Joyce S. Davis

First published 1979

ISBN: 0 7277 0079 0

© The Institution of Civil Engineers, 1978, 1979

All rights, including translation, reserved. Except for fair copying, no part of this publication may be reproduced, stored in a retrieval system, or transmitted in any form or by any means electronic, mechanical, photocopying, recording or otherwise, without the prior written permission of the Institution of Civil Engineers

The Institution of Civil Engineers does not accept responsibility for the statements made or for the opinions expressed in the following pages

Produced and distributed by Thomas Telford Ltd, P.O. Box 101, 26-34 Old Street, London EC1P 1JH

Made and printed in Great Britain by Inprint of Luton (Designers and Printers) Ltd

Contents

1 Ferrum corrumpitur

M. CLARKE, BSc, PhD, DSc(Eng), CChem, FRIC,
FIM, FICorrT, FIMF, Department of Metallurgy and
Materials, City of London Polytechnic

SYNOPSIS. An analysis of the origins of engineers' corrosion problems and the role of corrosion science, and of corrosion technology in their solution.

ENGINEERS AND CORROSION

1. Engineering is part of technology and one of the foundations of civilisation (ref.1). Technology is the use of tools, machines and processes to extend man's own faculties. Many branches are called engineering, but others, like medicine and agriculture, are not. Corrosion is an aspect of the decay of materials by chemical or biological agents. It may trouble a technologist in two ways: attacking his tools or machines, or spoiling his artifacts and annoying their users. In civil engineering the corrosion of artifacts is important - costly buildings, bridges, road and rail works, harbours, etc exposed to corrosion for long periods. Corrosion of civil engineering tools is a lesser problem. In chemical engineering however, corrosion of chemical plant is the main concern, while the artifacts - the chemicals produced - do not corrode.

2. Corrosion problems are created when the civil engineer decides to design using metal. The design adopted in relation to the metals selected, and the environment, fixes the nature of the problem. Unfortunately, often we do not know in advance either the problem or its solution.

DESIGN AND INVENTION

3. Engineering is creative and requires inspiration as do art, literature and music. The degree of novelty in the solution of an engineering problem covers a spectrum rang-

ing from <u>design</u> to <u>invention</u>. Civil engineering design consists of assembling parts, sub-systems etc., and specifying techniques for using them for a purpose. The individual steps and parts will be known and proved. The creative aspect is in the skill and economy with which they are combined. A problem may have several solutions differing in merit. An engineer's professional training aims to make him skilful in accepted design technique.

4. Invention produces a radical solution to a technological problem, with a novel departure from accepted design never seen previously. It discards experience. Professional designers are loath to solve problems by invention. They are trained experts in procedures which are 'correct', ie. already known, proved in use, judged good by their peers. A high risk attaches to novel solutions: for each successful invention hundreds fail. Yet men have an urge to invent, though the suppression of this by training often results in technologists inventing outside their own field (ref.2,3).

CREATING CORROSION PROBLEMS

5. Engineers create these by design and invention, though not deliberately. The corrosion problem is imported as a by product of intentions to achieve some other purpose. It is one of several matters subsidiary to an engineer's main concern, but this does not make corrosion easy to correct. Diagnosis of the corrosion problems associated with design in any field requires careful study of how the artifacts perform in service over fairly long periods of time. The results can, in principle, be incorporated in a manual of do's and don'ts rules, and of course this is done to an extent. But the information is diffused, often disregarded, and is not available in <u>advance of designs which have evolved and changed</u>.

6. Evans has said that one view of corrosion is as a disease from which metal can suffer (ref.4). This is a useful outlook for engineers. It follows that to appreciate corrosion problems one needs a clinical training, as in medicine, of 'walking the wards' diagnosing the corrosion disease from which metal objects suffer. This produces a view very different from that gained in the laboratory where corrosion is an artificial event in cells, beakers, or test tubes, stimulated or restrained to order. Passing on to invention, it is likely to entail new corrosion problems

which cannot be foreseen. Throughout history technological inventions have led to scientific discovery, but engineers do not rejoice when thus they discover new forms of corrosion. If a corrosion technologist could foresee trouble in a new invention and were consulted (which is unlikely) his gloomy views would tend to be brushed aside in the enthusiasm of innovation. The high optimism which greeted the Severn Road Bridge in the 1960's has not stood the test of time.

CAN ENGINEERING CORROSION BE PREVENTED?

7. Corrosion problems have not been eliminated from artifacts, but we may consider whether, in principle, they could be. The only conditions under which the answer would be 'yes' would be those of completely static design, because of the time needed to learn by experience whether corrosion were the factor limiting the life of the artifact. Corrosion itself need not be stopped: so long as an artifact's life is ended earlier by another factor, corrosion is tolerable. Railway track corrodes everywhere without major corrosion problems since it wears out by abrasion. However an invention which increased rail wear resistance one hundredfold while leaving the corrosion rate unchanged would create a corrosion problem. Even without invention, design evolution is likely continually to create new corrosion problems. It is worth noting that the avoidance of metals is a semantic solution to corrosion since non-metals are equally prone to decay by processes having other names. I have called this the Iron Law of Caducity: 'No matter what you make it from, it will not last forever'.

8. I conclude that while much can be done about engineer's corrosion problems, there is no permanent cure rendering technologists redundant, and enabling designers to relax their vigilance. The responsibility for mitigating corrosion should be shared. To expect that any and every design must have a solution to the corrosion problems it creates is unwarranted. We could draw up rules for current designs known for, say ten years or more, but not for designs yet to be conceived. Since the corrosion problem is a subsidiary factor, many engineers are, and will likely to continue to be, reluctant to make design subservient to the prescriptions of corrosion technology.

SCIENTIFIC AND TECHNOLOGICAL KNOWLEDGE

9. The term science is misused, often to cover both science as defined here, and technology as defined in paragraph 1. Here science is restricted to a special kind of knowledge, obtained by the method of hypothesis-deduction-observation. This science consists of theories, laws and explanations of phenomena, testable by inter-subjective observation or experiment, and accessible to anyone. Occultism, mysticism and private revelation is outside the realm of scientific knowledge. Natural phenomena and man-made inanimate phenomena are the province of physical science; social phenomena fall to social science to explain. The kind of knowledge produced by the scientific method can be called 'knowing-why'.

10. The form of technological knowledge is different, its purpose is different, and confusing it with scientific knowledge, though frequently done, is unhelpful. Technology is directed towards action for a purpose. The essence of technological knowledge is a set of directions and rules for the action: in the well-known phrase it is 'knowing-how'. The directions are variously called specifications, codes of practice, working instructions, recipes, etc, which we will subsume under the term operating principles. Knowing-how and knowing-why are different. Many are instructed how to ride a bicycle at an age when the theory of why they may defy gravity on two wheels is beyond understanding. Most cyclists continue in ignorance of theory throughout their career. Conversely, were bicycle theory mastered by a non-cyclist physicist, who doubts that if he forthwith mounted a machine devoid of know-how, he would bite the dust? Cycling know-how illustrates another difference: science can be, and is usually capable of being set down completely in writing and symbols - it can be recorded. Parts of know-how are extraordinarily difficult to record. I have never seen cycling know-how completely written down. It is invariably taught by one who has mastered it showing a learner how, and supervising his practice. Though much technological knowledge is written today, its correct interpretation still requires the possession of experience gained by a form of apprenticeship. The training of engineers recognises this. Polanyi calls the unwritable parts of know-how, 'tacit knowledge' (ref.5). Until the mid-nineteenth century practically all technological knowledge had to be passed on by showing since most people were illiterate (ref.6).

11. Technological knowledge is in the form of a series of imperative instructions for action. To bake a Christmas cake you turn to a recipe which tells you what you must do. The recipe contains negligible cake theory, ie. it does not explain why the recipe works, despite the fact that this technology is often taught as Domestic Science. Science pre-supposes no purpose in its students apart from curiosity. Neither the theory of relativity nor the theory of galvanic corrosion tell you to do anything: they are not technological knowledge. While know-how has existed independently of records, written symbols are inseparable from all but the most rudimentary science. It is the aim of science to be accessible to all who will learn, and a social rule of scientists is to disregard argument from authority. Technology must be different. When it comes to doing something - cycling, baking cakes or designing bridges - the master who knows speaks with authority to the learner who does not. Rational argument, and nowadays science, may be used to support operating principles but in know-how there is always an implication 'I know how, and if you wish to know how as well, you should do as I say'. It is this proposition which renders the inventor so lonely and exposed, since he has to claim 'I, and only I, know how to do this ...'. Such messianic personal revelation always arouses scepticism amongst the experts. In the end however the independent proof of a new operating principle is that it works, and can eventually be passed on to others. No matter how 'illogical' or 'unscientific' in the light of accepted theory, if a new operating principle works, it is the science which must give way, not the invention.

12. Part of physical science is theories which explain technology. Inventions often present science with puzzles. Savery and Newcomen perfected a steam engine in the period 1695-1710. Philosophers puzzled why it created energy, since heat was not then held to be energy. Carnot's 'Refections on the motive power of fire', of 1824 explained matters and founded the science of thermodynamics. Appert invented food preservation by sterilisation in a sealed container in 1795. This is the basis of bottling and canning food. The science of microbiology which explained the process was due to Pasteur in 1861. Volta invented his 'pile', the first electric battery, in about 1800. The science of electrochemistry, which also explains much corrosion, followed, as did electrical engineering. A science which is founded on an invention (something never

before seen) can only follow the invention. It is false
to say the steam engine is a product of applied science,
despite the all-too-common view that inventions flow from
science. It may be the fertility of science in guiding
imitation and analogy in certain fields which gives rise
to the latter view. Chemistry is very useful for 'inventing' new drugs or new plastics, once there exists a proto-
type, by guiding the synthesis of similar molecular structures with a reasonable hope of profitable success. While
requiring considerable skill, this process is really at the
design end of the spectrum, and the 'reasonable hope' is
still a piece of faith, not a cast-iron part of scientific
logic. The history of technology (ref.1) and recent studies of invention (ref.2, 3) refute the notion of technology
as applied science. Technology may go far ahead of science
and has managed well with no science at all both in the
west (ref.1) and the east (ref. 7).

CORROSION KNOWLEDGE AND THE PROBLEM OF COMMUNICATION

13. In 1971, a committee appointed by the Minister of
Technology, estimated the national cost of metallic corrosion (ref.8). It concluded that about a quarter, £310x10^6,
1971 prices, could be saved by the effective use of existing knowledge. Eight years later the Ministry of Technology has vanished, but not corrosion. I doubt that the
savings have been made since the knowledge on which such
store was set is corrosion science, not corrosion control
technology. We know a good deal about why corrosion occurs
from the research of a few pioneers and numerous scientists
more recently. Their objectives were to understand the
causes and mechanisms of corrosion processes. The phenomenon of the corrosion of metal artifacts (and all metal
except gold occurs as artifacts) has been known for centuries, but was explained only when it was moved into laboratories and studied by scientific methods. But the result
is corrosion science, and to corrosion scientists corrosion
is a fascinating laboratory puzzle to be explained, rather
than a disease suffered by engineering structures to be
cured.

14. Explanations of corrosion are not operating principles
for controlling it. Corrosion science can be useful to
those trying to formulate anti-corrosion recipes, but not
necessarily so. The gap between explaining something and
being able to do anything about it is analysed factually by
Jewkes, Sawyer and Stillerman (ref.2). Design or invention

is needed to formulate corrosion control techniques, and corrosion science no more provides this effortlessly than does physics design bridges. Physics says nothing of bridges as such, though a knowledge of physics might enable a designer to avoid unstable structures without experience teaching him that they fall down. But bridges built for centuries without physics stand yet, while innovative modern designs apparently reached ahead of physics and fell(ref.9). Preaching corrosion science at engineers is undoubtedly seen by many as a major contribution, and when it fails to inspire them to solve their problems, this in turn is dubbed 'a problem of communication'. The published discussion of the 1971 report (ref.10) reflects this:-

"... it was really difficult to get scientists to talk sense in the engineer's way, and it was equally difficult to get engineers to listen."

"... there was a psychological block."

"... upper crust and shop floor should get together to discuss problems."

A strong, though not unanimous, view was expressed by corrosion experts that the problem was one of educating engineers. This view persists, so that a 'Design Guide' which appeared in 1977 is still geared more to scientific explanation than to rules and recipes for stopping corrosion (ref.11). There is a problem of communication but it is that of conveying to corrosion experts the difference between scientific and technological knowledge, and the need for more of the latter kind.

INDEPENDENCE OF OPERATING PRINCIPLES AND SCIENCE

15. Some participants to the 1971 discussion pressed the engineer's need for specifications, but others claimed pre-eminence for science over operating principles.

"... specifications have their limitations. Things changed over the years and we should be careful not to give them an overall blanket."

"... specifications without explanation could lead to the discrediting of corrosion education or information centres. Fundamental information must be got to those who operate ... "

The Chairman of the 1971 Committee, Dr Hoar, restated his views in 1977 (ref.12):-

7

"... I don't believe there is any intellectural difference between the scientist and the technologist. The
technologist is merely somebody who uses his scientific
knowledge to solve practical problems ... Let us get
away from this theme that some of us are practical men
and others are theoretical men: we all ought to be about
50/50 and then the advance can be greater on all
fronts ..."

Through analysis of engineering corrosion problems, I have
moved to a view diametrically opposed to the consensus of
these quotations (ref.13), which I believe perpetuate a
confusion. The fact is that theory (science) and practice
(operating principles) are logically independent. No operating principle has ever ceased to work because the theoretical explanation of it has been overthrown. Fires burned
unaltered when the phlogiston theory was refuted. Devices
functioned unchecked when Newton's mechanics, on which they
were 'based', was overthrown by Einstein. Telescopes
worked on as theories of light executed u-turns from corp-
uscles to waves and back to photons. Radio communication,
invented to apply the physical propagation of electromag-
netic waves through the ether, was not silenced when phys-
icists admitted there was no ether. Now if operating principles were merely 'applied science' this would need some
profound explanation.

16. The title of this paper is culled from a much earlier
writer, Gaius Plinius Secundus, AD23-79. His 'Natural
History of the World' contains much technological knowledge (ref.14). Book 34 deals with 'Ferrum ...corrumpitur'
or spoiled iron. Corrosion was troublesome then as now,
but Pliny records:

"It can be protected from rust by means of white lead,
gypsum and vegetable pitch."

This is a sound recipe, and cannot possibly be applied
corrosion science, which it precedes by eighteen centuries.
Pliny was not the inventor, but the recorder of something
well-known, and which in similar forms continued to be
used by practical men to the present (ref.13,15). Lacking
the scientific method, Pliny nevertheless asked, as would a
scientist: why should iron corrode more easily than other
metals? This is still a good question. His answer is that
it is because iron is both the best and worst of man's
servants. Very useful domestically, it is also the metal
of war, slaughter and brigandage. Of iron arrows: it is

a great evil that 'to enable death to reach human beings
more quickly we have taught iron how to fly and have given
wings to it." Corrosion compensates: 'The same benevolence
of nature has limited the power of iron itself by inflict-
ing on it the penalty of rust, and the same foresight by
making nothing in the world more mortal than that which is
most hostile to mortality.' Pliny's metaphysical theory
would support a second technique he reports, corrosion control by a religious ceremony. It was used for the chains
of a suspension bridge built for Alexander the Great but
Pliny is sceptical since the method failed everywhere else.
But the test for technology is not the truth of any science
'underlying' it, but the answer to the question: 'Does it
work?.' The pitch recipe works today, as in Roman days,
though corrosion science has changed beyond recognition.
Philosophers of science following Popper (ref.16), find
the most that logic can say about a scientific theory is
that it is not yet refuted. At any future time it may be;
there can be no absolute verification of theory. Conversely, history suggests that operating principles can be
verified absolutely by the test 'does it work?' They cannot
be refuted absolutely, though they may be forgotten. Technological knowledge is not scientific knowledge.

CONCLUSIONS

17. Civil engineers with corrosion problems need operating
principles to tell them what to do to cure the trouble.
Some excellent recipes, codes of practice and specifications
exist, emanating from various sources. As with most technological knowledge, an engineering type of training and
outlook (including the 'tacit knowledge' of the art) is
needed to make best use of anti-corrosion technology. In-
dividuals, who may draw their inspiration from corrosion
science or from listening to Mozart, will have to study
the problems and invent or design measures to solve them.
Since corrosion problems follow in the wake of other
engineering innovations, it seems inevitable that corrosion
control operating principles will have the flavour of
fighting a previous war. The corrosion scientist's cont-
empt for operating principles, which he often dismisses as
'a cook-book approach', needs correction. If devising new
successful recipes required so little talent we would have
no problems. This to my mind is the real 'psychological
block' to be overcome.

CORROSION IN CIVIL ENGINEERING

NOTE

18. In this paper I have, intentionally, concentrated on corrosion problems as being those for which existing control techniques do not work. I have not been concerned with corrosion arising from ignorance, or neglect, of techniques which could correct it.

REFERENCES

1. SINGER C et al (Ed). A history of technology. OUP, Oxford, 1954-8.
2. JEWKES J.,SAWYER D., and STILLERMAN R. The sources of invention. 2nd Ed. Macmillan, London, 1969.
3. LANGRISH J et al. Wealth from knowledge. Macmillan, London, 1972.
4. EVANS U R. The corrosion and oxidation of metals. Arnold, London, 1960, 16.
5. POLANYI M. Personal knowledge. Routledge, London, 1958.
6. ARMYTAGE W H G. A social history of engineering. 4th Ed. Faber, London, 1976.
7. RONAN C A. The shorter science and civilisation in China. Vol.1. CUP, Cambridge, 1978.
8. Report of the committee on corrosion and protection. HMSO, London, 1971.
9. Report into the failure of Westgate bridge. State of Victoria, Melbourne, 1971.
10. Corrosion and Protection. Institution of Mechanical Engineers, London, 1972.
11. ROSS T K. Metal corrosion: engineering design guide No.21. OUP, London, 1977.
12. HOAR T P. U R Evans award lecture. Bulletin, Institution of Corrosion Science and Technology. No.66, 1977, 6-11.
13. CLARKE M. The relation between theory and practice. British Corrosion Journal 1976, 11, 113-6.
14. PLINY. Natural history of the world. Heinemann, London, 1938-62, book 34.
15. CLARKE M. Scientific and historical explanations and their place in technology. British Corrosion Journal 1978, 13, 1-4.
16. POPPER K R. The logic of scientific discovery. Hutchinson, London, 1959.

2 Corrosion protection as seen by an engineer in a large organisation

D. F. GOODMAN, BEng, MICE, MICorrT, British
Railways, Marylebone

SYNOPSIS. The selection of a surface coating requires a realistic consideration of all the factors
affecting the coating, both chemical and non-chemical, and including future maintenance where required. This paper deals, in the main, with the
non-chemical side. The problems are illustrated
by the example of some railway bridges and attention is drawn to the importance of local effects
and of actual conditions which cannot always be
forecast.

INTRODUCTION

1. When protection is required for a long life
two distinct but related activities are involved:
firstly the provision of a suitable coating and
secondly the subsequent maintenance of this coating throughout the life of the structure. These
two activities must be considered as a whole.

2. The provision of such protection raises a number of problems which can only be resolved in the
light of conditions and knowledge at the time.
Some of these problems are of an engineering/orga-
nisational nature and some of a surface coating
nature.

3. The many factors involved and the changes that
take place with time mean that, taken overall, few
solutions completely dispose of problems. However
thorough consideration in advance can help to
minimise them.

4. The following comments are generally restricted to bridges. However, many of the problems are

the same for other structures although the conditions of use may vary.

5. It will be appreciated that the factors concerned in the railway industry may or may not apply to other industries.

TREATMENT OF NEW WORK

6. In trying to decide which treatment is best it is necessary to take into account all the potential problems affecting each stage of the process including future maintenance where relevant. Many of these will undoubtedly come readily to mind, for example:-

(a) The possibility of incorporating anti-corrosion features in the design.

(b) The adequacy of the facilities at the fabrication shops to cope with adverse environmental conditions.

(c) The susceptibility of coating systems to damage during transit and erection.

(d) The reliability of the systems available in the situations involved.

(e) The maintain-ability of the systems under consideration in the sense that breakdown proceeds in such a way that it is obvious before it becomes complete and can be dealt with so as to prolong the life of the system.

(f) Compliance with Health and Safety requirements.

Others which may not be so obvious can be equally important such as:-

(g) The effect on the systems of any repair work in the future including Health and Safety aspects.

(h) The effect of possible changes in the site environment with time.

(i) The conditions under which maintenance will be carried out. This is closely linked with (d) above.

(j) The effect of changes in materials, systems and techniques on the staff concerned with future maintenance.

7. In the case of item (d) the intending specifier is presented with a bewildering range of materials and systems. A good laboratory whose staff have adequate knowledge and experience can tell if a material/system is intrinsically good or bad but cannot anticipate all conditions likely to occur in practice so that the final test can only be that of time.

8. With regard to items (e) and (i), when a long life is required the system selected obviously needs to be maintainable. It is not however always realised that the conditions under which much maintenance has to be carried out makes it difficult to achieve satisfactory standards of work, particularly when corrosion has started. Maintenance frequently involves taking possession of the area around the structure which can interfere with other activities and these activities in turn can limit the time available for the work. On B.R. the railway lines and frequently the overhead wires are involved which affect the train service and therefore possessions have to be kept to the minimum.

9. Another point which must be borne in mind is that the maintenance of surface coatings is particularly subject to the availability of resources so that there can be no guarantee that maintenance will be carried out when the need first becomes apparent.

ADVICE AND INFORMATION

10. The information required resides in a number of people whose interests are not necessarily the same and whose knowledge and experience lie in different fields and this raises the first problem - how to obtain adequate information about all these aspects.

11. When, some years ago, the Civil Engineering Department of B.R. decided on a major revision of its recommendations and procedures to take account of modern trends and developments, this particular problem was dealt with by forming a group comprising persons with experience as Civil Engineers, Supervisors/Inspectors, Paint Technologists and

Purchasers together with a Mechanical & Electrical Engineer and an Architect.

12. The group, which has a continuing existence, strives to produce solutions which are of maximum overall benefit to B.R. It will be appreciated that this arrangement is not open to many organisations from within their own ranks.

SELECTION OF MATERIALS/SYSTEMS FOR NEW WORK

13. The work necessary to take account of all the problems can be illustrated by B.R.'s bridges. The overall problem could be said to be the selection of a minimum number of materials/systems tolerant of the adverse factors which may be encountered at the various stages of fabrication, erection and in use, so as to give a reliably long life to the time when the initial system first needs maintenance for the majority of bridges in the majority of geographical situations in the British Isles; also to be maintainable within the constraints.

14. The general requirement is that these structures will do what is needed, safely, at minimum cost and with acceptable appearance throughout their lives, which tend to range from about 25 years to in excess of 100 years for the majority of bridges.

15. The endeavours to solve this problem led to an investigation in which the characteristics of the systems used by, or thought likely to be useful to B.R. were considered.

After some preliminary thinning out the following systems were examined:-

Paint Systems - Micaceous Iron Oxide; Chlorinated Rubber; Two Pack Epoxy Based Types.

Metallic Systems - Hot dip galvanising without paint; Hot dip galvanising plus paint; Zinc spray plus paint; Aluminium spray plus paint.

Weathering Steel - Corten "B"

16. For the purpose of the exercise laboratory comments were obtained and information sought from both British Rail and outside sources as to

the characteristics of these materials/systems and in particular to:-

(a) Application. Amount of work which has to be or can be carried out at the steel mill or fabrication shops and the advantages, disadvantages and problems of doing so; also the problems of site application. It was felt that in general it was advisable to have systems in which the maximum amount of work could be carried out at the fabricators. This because the conditions at fabrication shops should be more favourable to the achievements of satisfactory standards than those at site.

(b) Use of high strength friction grip bolts. In view of the use of h.s.f.g. bolts on B.R., the possibility of developing suitable slip factors with the systems under consideration when applied to faying surfaces; also the possibility of treating such systems to obtain suitable slip factors.

(c) Chemical attack and abrasion. The effect of fuel oil, petrol, road salt, limestone and granite ballast as substances likely to come into contact with surface coatings alongside the track.

(d) Transport and erection. The resistance of the systems to damage during transport and erection.

(e) Compatibility. The compatibility of each of the systems with existing maintenance systems.

(f) Limitations on use. Any limitations on use in particular situations.

(g) Life to condition where repainting desirable. The maximum life actually obtained or, if not yet reached based on experience to date, in areas of high and low pollution. For this purpose experience of the actual behaviour of structures on B.R. was drawn on and large numbers examined as were a number of non-railway structures.

(h) Comparative cost. Based on (g) the comparative cost of protection was calculated for a specific structure on the basis of the likely range of costs of initial coating and subsequent maintenance.

17. It was not possible to examine equal numbers of all types so that the exercise was not statistically as satisfactory as could have been wished. Nor was it possible to obtain full details of the actual histories of each structure seen. The approach represented the best possible in the circumstances. However, the picture emerged so clearly that there were no qualms about the conclusions.

RESULTS OF SITE INVESTIGATION

18. Information in respect of most of the items from (a) to (f) in para.16 can be obtained by anyone prepared to spend the time necessary to go through the literature on the subject. It is not intended to dwell on these nor to cover the behaviour of the systems considered in detail. It is however interesting to note the broad picture revealed by item (g) para.16, which may be summed up as follows:-

Micaceous Iron Oxide (M.I.O.). Systems originally used consisted of lead based primers and two coats of M.I.O. over grit blasted surfaces with zinc rich blast primers. More recently the lead was replaced by a zinc phosphate primer.

It was found that whilst exposed surfaces had a life in the order of 10 years to the condition for first maintenance the time was reduced to about 2 years, in areas subject to condensation or damp, broadly speaking irrespective of the actual system applied.

Chlorinated Rubber Paint (C.R.P.). The few structures considered indicated confirmation of the laboratory view that the material was no better than M.I.O. in "good" areas but better, thickness for thickness, in "poor" areas.

Two Pack Epoxy Based Type. The sensitivity of early formulations of this type of material to application conditions and overcoating time was reflected on site. Numerous cases being found of intercoat adhesion failure. Recent formulations appear to be less sensitive but there is no long term site experience as yet.

Hot Dip Galvanising plus Paint. Experience mainly with overhead electrification gantries erected in the last years of steam traction in which the system consisted of 2 oz. zinc per sq.ft. with either one coat of calcium plumbate and two coats of M.I.O. or one coat of single pack etch primer and 10 thou. of gel. bitumen applied after weathering. Generally 15 to 20 years life to need for first maintenance in mildly polluted environments dropping to about 10 years in some heavily polluted areas.

Hot Dip Galvanising without paint. Experience in the main with overhead electric gantries protected with a minimum coating of 2 oz. per sq.ft. of zinc. Life to need for first maintenance generally in excess of 20 years in rural areas but less in industrial areas and those subject to local pollution. Markedly different rates of deterioration could be seen on different sides of the same structure, presumably due to the effects of air currents.

Metal Spray plus Paint. Examination of railway bridges treated with both zinc and aluminium spray and overcoated with paint showed these to be in very good shape with lives to the condition for first maintenance from about 10 to more than 20 years for both zinc and aluminium. Paint treatment for the older structures was zinc chromate or red lead and M.I.O., with the addition of a white decorative paint where light reflectance required, more recently zinc phosphate and C.R.P.

Deterioration of the paint followed the same pattern as with non-sprayed structures with the first breakdown in condensation areas. However, when such deterioration took place the existence of the spray provided protection for many years when maintenance was delayed. The areas concerned were invariably quite small. Of the two metals, aluminium seemed marginally to give the better protection.

Weathering Steel.

A number of foot and road bridges over the railway and canals were inspected. Whilst the majority of the steelwork appeared to be developing a satis-

factory patina, small areas over piers or abut-
ments, where there was leakage and often restricted ventilation, seemed to be rusting with the development of normal corrosion.

19. Besides the general performance of the systems
the site visits confirmed the importance of several facts, firstly that in time structures may
develop defects other than those associated with
surface coatings, secondly, that site conditions
may present severe limitations on what can be done
by way of maintenance painting when this is needed
and thirdly that actual conditions can produce
severe localised corrosion in otherwise sound
structures.

20. In all cases the most rapid deterioration was
found on those parts of the structure subject to
condensation, and which were often badly ventilated, those parts subject to leakage or splash
and those parts in contact with wet material for
long periods.

21. In the case of footbridges examples abounded
in which material such as paper, leaves and soil
would lodge between protective screens and structural members. This would remain damp for long
periods holding moisture against the steelwork
leading to rapid breakdown of surface protection
and, in the worst cases, to severe corrosion of
steelwork.

22. A similar situation with rail bridges involved
ballast and general rubbish lodged against steelwork, often in the narrow spaces behind pilasters.
In road bridges also rubbish, soil and grit, had
built up at the bottom of webs adjacent to paved
surfaces, bringing about the same effect.

23. Another area prone to localised attack was the
bottom of columns. Here the conditions also tend
to produce long standing damp material in contact
with the steelwork even without the existence of
debris.

24. With regards to leakage water seems eventually
to find its way behind a lot of bearings where,
because such places tend to be badly ventilated,
the area remains damp for long periods. Water
also percolates past waterproofing, penetrates

expansion or other joints and overflows gutters when normal drainage becomes blocked.

25. In a number of cases, maintenance painting had been rendered ineffective by the fact that it had been carried out without regard for some obvious defect such as leakage.

26. Where repainting had been left too long and corrosion obtained a hold, the maintenance treatments possible in practice at the time often failed to stop the corrosion. In such cases the rate of breakdown of the maintenance paint was rapid. No such metallic system seen, however, had reached this condition.

27. One feature noticed throughout was the existence, marked in places, of physical damage to coatings. This appeared to be due to the top flanges of some bridges being used as walkways, to top flanges and inner webs being hit by track maintenance tools or ballast and to external surfaces being hit by missiles thrown by children or to being kicked. On non-metallic coated structures this produced failures which tended to lead to corrosion. On metal sprayed structures the failure took place over the spray coat and remarkably little rust was found.

DISCUSSION OF RESULTS

28. The tendency of M.I.O. to early failure in condensation areas was obviously unsatisfactory in relation to the long life requirement and maintenance constraints as was that of C.R.P. Although epoxy pitch looked more promising, its dark colour was a limitation in most situations.

Other epoxy based systems seen were either not sufficiently tolerant of the conditions or of too recent introduction to enable conclusions to be drawn. The apparent tendency to corrosion of weathering steel in areas usually subject to adverse conditions in railway bridges did not give the requisite degree of confidence in this material for these situations.

29. Hot dip galvanising plus paint performed well but with some limitations as to faying surfaces and size of members. Heavy hot dip galvanising

looked attractive but there was no example of long
term behaviour in areas of high pollution. Metal
spray plus paint much more nearly met the conditions and the calculations of cost indicated that
it compared favourably with other systems in the
long term.

30. It was concluded that one of the reasons for
the long life to first maintenance of the metallic
systems was that they seemed to cope better with
the problems of fabrication, transport and erection so that they commenced their life on site
substantially intact. This is of course particularly important where there may be difficulties or
delays in making good or overcoating. The final
paint coats are then applied to reasonably sound
surfaces and so the paint system is helped to
attain maximum life.

31. A final paint system is considered to be
necessary partly because of the desire for uniform
appearance but principally because of the requirement of maintainability. With an overcoat of
paint the need for maintenance can be seen when a
base coat, say the sealer coat in the case of
spray, is revealed. If all is well the paint coat
can be kept reasonably intact with minimum effort
but if the work is badly delayed many years of
protection are left in the metallic coats.

32. Of the metal coatings the performance of aluminium spray was particularly fortunate as it
develops slip factors considered satisfactory for
friction grip bolted connections in railway bridges. However, it is an expensive system initially
and cost is also an important factor. The material is not foolproof and there have been cases of
unsatisfactory application.

33. With regards to maintenance, it is clear that
if the life of the structure is determined by some
feature, for example leakage, which cannot be
treated, or corrosion in some inaccessible place,
then the justification for any maintenance painting needs carefully reviewing.

Further, unless it is possible to prepare surfaces,
particularly corroded surfaces, to a satisfactory
standard and apply an adequate coating under the

prevailing site conditions, the same comment applies.

34. Perhaps the most interesting feature to emerge was the importance of local effects. The selection of the best system is not the end of the problem. Great attention must be paid to the design and treatment of problem areas. After all it does not matter if the chosen system is twice as good as some other if failure results from unchecked corrosion at a critical point with which neither system can cope.

ACKNOWLEDGEMENTS

I would like to express my thanks to the Chief Civil Engineer of the British Railways Board - Mr. M.C. Purbrick for his permission to produce this paper and to the members of the Surface Coating Sub-Committee and other colleagues who assisted in its preparation.

THE COPYRIGHT OF THIS PAPER BELONGS TO THE BRITISH RAILWAYS BOARD.

3 BS 5493: Code of practice for protective coating of iron and steel structures

K. A. CHANDLER, BSc, FICorrT, British Steel Corporation

SYNOPSIS. BS 5493 is considered in relation to the economic protection of steel structures against corrosion. Some of the considerations influencing the drafting committee's views are discussed and general comments are made on the various sections of the code.

INTRODUCTION

B.S. 5493 was published at the end of 1977 to supersede CP2008, originally published in 1966. CP2008, was the first code of practice for the protection of structural steelwork produced in the United Kingdom. Despite the undoubted achievement of providing a document of this nature it soon became subject to criticism; some of it justified but much of it misplaced. In part the criticism arose from a misunderstanding of the nature and purpose of the Code.

Similar criticism will occur with B.S. 5493. This is to be expected because complete agreement on a document of this type is unlikely to be achieved. The various parties involved in the protection of steelwork have different interests and requirements. There is nothing unusual about this. In most industrial situations the same problems arise. Consumers are critical of manufacturers, who in turn may be unhappy with their suppliers. In the final analysis, the customer, i.e. the purchaser of the goods or service, has to be satisfied, but to achieve this objective all concerned must be working towards this end. This involves a considerable interplay of interests. Steel interests clearly want structures to be in steel but equally coating suppliers

depend upon constructions that require coatings. There
is, of course, competitiion between different types of
coating, e.g. metallic and organic, and within each of
the coating groups. The customer has to find his way
through a fairly complex situation and a good deal of
misunderstanding can arise in the process. The customer
often adds to the problems by not providing his
requirements in a precise way.

One method of overcoming these problems is to prepare
standards, specifications and codes of practice so that
all those concerned in a project are speaking the same
language and can, within reasonable limits, make valid
comparisons.

The aim of those preparing B.S. 5493 was to provide a
document that could be used in this way. The emphasis
was placed on satisfying the final customer because as
noted above, he is the key to the whole operation.

In the process of preparing the Standard there were
inevitable conflicts of interest. These were generally
resolved by discussion and co-operation but in certain
areas minority opinions could not necessarily be
satisfied. Furtheremore, there has to be a limit on
deliberations if a Standard is to be produced within a
reasonable time-limit. Consequently, improvements can
always be made particularly with the experience gained in
the use of a standard. It follows that all standards
should be under scrutiny and require revision at
reasonable intervals. There are practical difficulties
in carrying out revisions at too frequent intervals.
British Standards, however, is adopting a policy of
reviewing its publications at regular intervals and it
would be expected that B.S. 5493 would, in the not too
distant future, be considered for revision and updating.

2. <u>Coating specifications</u>. In the field of steel
protection, the most widely used material apart from
steel, is paint. A number of standards for paint exist
but there are few British national standards in this
area. It is not my intention to discuss this matter in
any detail. Considerable problems arise when considering
paint standards. I do not believe that they are
insurmountable but they require careful consideration.
The basic argument concerns compositional as opposed to

performance standards. Paint manufacturers on the whole feel that performance should be the criterion to be adopted in judging their products and that they should, within reason, be free to develop and formulate paints in their own way. There are powerful arguments to support this view but the matter is not as simple as might be supposed. If the term 'Performance' is to take into account all the requirements of the user such as life to first maintenance, ease of application and colour retention, then it is a sound criterion for judging paints. Problems however arise and these can be illustrated by considering the 'life' of the paint. In practice the interest is in the life of a system rather than that of individual paint films. Furthermore, other factors, in particular surface preparation and application have a significant effect on performance.

Taking all these factors into account the Drafting Committee for B.S. 5493 decided that performance require-ments for paints could not be included in a code of practice. Nevertheless, it seemed desirable to provide users with some method of characterising paints. The Paint Industry co-operated with the committee and provided three characterstics for the paints mentioned in the standard. These were:-

Volume solids (nominal %)
Main pigment in total pigment (weight %)
Dry film thickness.

Additionally the binder and main pigment were noted for each paint.

These characterstics provide no more than a very broad indication of the type of paint. Certainly, they cannot be considered as being a formulation or paint composition. There is some truth in the assertion that such information can be misleading if used incorrectly. This was considered by the drafting committee but the concensus view was that the information, albeit limited, would be of benefit to both users and paintmakers. To specify paints in vague terms such as Micaceous Iron Oxide or Zinc Phosphate is not satisfactory. Although paint manufacturers generally provide additional informa-tion in their technical literature, it is not always easy with the use of proprietary names to be clear whether

Table 1.   Exterior exposed non-polluted coastal atmosphere(based
on BS 5493 Table 3 Part 4).

Typical time to first maintenance (years)	General description	System reference (table 4)	Total nominal thickness (μm)
Very long (20 or more)	Galvanize	SB2	140
	Unsealed sprayed aluminium	SC2A	150
	Unsealed sprayed zinc	SC3Z	250
	Sealed sprayed aluminium	SC6A	150
	Sealed sprayed zinc	SC6Z	150
Long (10 to 20)	Galvanize	SB1	(85 min.)
	Galvanize plus paint	SB9	(85 min. + 60 min.)
	Unsealed sprayed zinc	SC2Z	150
	Sealed sprayed aluminium	SC5A	100
	Sealed sprayed zinc	SC5Z	100
	Sprayed aluminium plus paint	SC9A	100 + (30 to 100)
	Sprayed zinc plus paint	SC9Z	100 + (30 to 100)
	Organic zinc-rich	SD3	100
	Inorganic zinc-rich	SE2	100
	Drying-oil type	SF8	190 to 230
	Silicone alkyd over two-pack chemical-resistant	SG1	245
	One-pack chemical-resistant	SH6	270
	One-pack chemical-resistant over two-pack chemical-resistant	SL3	295
Medium (5 to 10)	Unsealed sprayed zinc	SC1Z	100
	Organic zinc-rich	SD2	75
	Inorganic zinc-rich	SE1	75
	Drying-oil type	SF7	165 to 190
	One-pack chemical-resistant	SH3	150
Short (less than 5)	Organic zinc-rich	SD1	50
	Drying-oil type	SF2	120 to 150
	Drying-oil type	SF5	85 to 105
	One-pack chemical-resistant	SH1	160

competitive products are of the same general type.

The whole question of paint specifications is one to which attention is at present being paid in various quarters. It is a complex matter but an improved method of designating paints would be expected to evolve.

Metallic coatings make up the other major group of protective coatings. There is a wide range of standards covering them and as on the whole they are basically simpler materials than those in the organic group, fewer problems exist in this area.

3. Coating performance. In Section Two of B.S. 5493, a series of tables provides information for typical periods to first maintenance for a wide range of protective systems. A considerable amount of thought was given to the provision of these tables. Clearly, it is difficult to provide technical information of this kind for coatings that may be applied under a variety of conditions. Furthermore, although the information is tabulated according to the general environment, inevitably local conditions will have a significant effect on performance.

Systems are tabulated for ten main environments and notes are provided for a further nine. Table 1 indicates how these tables were compiled. The information in the B.S. 5493 tables was supplied by the suppliers of the various coating materials. One result of this type of tabulation is to emphasise the long-term performance of metal coatings. For very long periods, there is little doubt that properly applied metal coatings will, in many environments, provide longer lives to first maintenance than will paints. The tables do not, of course, take into account the probability of achieving the lives given. Factors, such as sound application procedures, correct storage and proper handling will determine the degree of success attained. The information provided does, however, indicate the typical lives to be expected when work is carried out in accordance with all the provisions of the standard. This is an important point – the tables can be used only as part of the overall recommendations of the Code and must take into account the other sections, particularly specifications and inspection.

4. <u>Specifications and technical requirements</u>. Many steel users still do not fully appreciate the importance of specifications. The performance of paint coatings in particular are very much influenced by the control of application. The paint in the can is the material used to provide the dry paint film but it should not be assumed that the same paint will necessarily provide films that will perform iin a similar manner. The manufacturer's instructions must be followed and for work of any importance a specification must be provided. The term specification as used in the code denotes the means of communicating requirements regarding the quality of materials and standards of workmanship necessary to provide good protection to steel structure'. Advice on preparing technical specifications is provided in B.S. 5493. Although in practice a specification may be both a technical and contract document, the essential feature – that of communicating the requirements of the work – is not affected.

The Drafting Committee considered the possibility of preparing 'model specifications' but this proved to be unrealistic in view of the large number of possible variations in the requirements for different constructions. Some steel users may take the view that the preparation of detailed specifications is unnecessary and that they would be in a stronger position if they based their requirements on performance. This may be a valid view in some situations where clear guarantees are provided. On the whole, though, guarantees of protective systems do not appear to have been as successful as might have been expected. Even where guarantees are used, a concise specification must be prepared by the guarantor to safeguard his own interests.

5. <u>Inspection</u>. The quality control aspects of pro-tective coatings are covered in the Inspection section of B.S. 5493. Inspection is basically a method of ensuring that the specification is adhered to. It is an essential part of coatings application. Although inspection is widely carried out by specialist firms, it is the responsibility of all parties engaged in the protection of steelwork to ensure that the work is carried out in a proper technical manner. The standard contains a guide to defects that can arise during protective coating operations together with likely causes and suggestions for action.

Advice is given on the scheduling and recording of inspection work. Users often comment on the lack of a nationally recognised certificate of competence for inspectors. This matter is at present under review and some national scheme may be developed.

6. <u>Maintenance</u>. A section on maintenance is included in B.S. 5493. It is, perhaps, surprising that only about 10 per cent of the total code deals with this aspect. The main emphasis is on the initial coating of steelwork and the importance of starting with a good system, well applied, requires no emphasis. Nevertheless, on most structures more money is spent on maintenance painting than on the original protection. It could be said that insufficient attention is paid to maintenance e.g. there are few coatings specially formulated for this purpose. To some extent the Standard reflects this situation. On the other hand it is made clear in the Code that considerable savings are possible if maintenance painting is carried out to the highest possible standard before undue deterioration of the coating has occurrred.

7. <u>General discussion</u>. The economic protection of structural steelwork against corrosion is not a simple matter. Most of those ultimately responsible for the protection of steelwork are not specialists in coatings and corrosion, so the guidance of B.S. 5493 should prove useful provided it is studied with some care. Advice can and should be sought by the non-specialist but a good deal of this advice is naturally biased towards certain products. With the assistance of the Code he is in a much better position to judge the validity of the advice offered. The majority of firms engaged in the supply of materials and services in the protective coatings field offer useful advice and are often the most qualified to provide the information required. Their expertise should not be ignored but responsibility rests with the customer and he must make the ultimate decisions.

Health and Safety requirements are becoming increasingly important and it is, of course, the duty of users of the code to ensure that all statutory requirements are met. It is beyond the scope of the code to discuss these matters in detail but a short section is included. This draws attention to typical aspects that arise during the process of coating steelwork.

CORROSION IN CIVIL ENGINEERING

The influence of design on the performance of coatings is discussed in an appendix to the code. This is a matter to which designers must pay more attention than has been the tendency in the past and typical points are discussed.

4 Protection of structural steel work

Some U.S. experience and practice

J. D. KEANE, Director of Research, Steel Structures
Painting Council; Senior Fellow and Manager,
Carnegie-Mellon Institute of Research

I am honored by your invitation, which I received through
Dr Bishop and Ms Brooks. Actually the subject of this conference was first broached to me by my good friend, the late
Julyan Day. All of us were touched by his personality. I
myself have had the deepest admiration for his indomitable
love of life.

The theme you have chosen for your conference is an important one throughout the world. Metallic corrosion costs in
the United States, for example, have recently been estimated
at 70 billion dollars per year or 4.2% of the estimated
gross national product. Of this total about 15% or 10 billion dollars was deemed to be avoidable. This compares with
an earlier estimate by Mr Hoar's committee setting corrosion
costs in the United Kingdom for 1969-70 at about 3.5% of the
gross national product with 23% of this total considered
potentially avoidable. Similar estimates in Russia, Germany,
Finland, Sweden, India, Australia, and Japan clearly identify
avoidable metallic corrosion as a world-wide problem deser-
ving high priority both in application of established technology and in new research and development.

Indeed the painting of structural steel could be described as the principal means of protecting our principal
raw material - steel - against its principal limitation -
corrosion. In this mission the prestigious BS5493 is a monu-
mental guide quite distinctive from anything in the U.S. or
elsewhere. Such codes of practice are designed to aid the
civil engineer or other specifier, who is concerned with
many other factors as well, in his responsibilities for
specifying optimized anti-corrosion schemes.

I have been asked to provide some examples from the specifications of the Steel Structures Painting Council (SSPC),
since these are the ones with which I am most familiar.
Incidentally the SSPC specifications and manual, long in use
in the U.S., are once again being reviewed and renewed, a
process in which your participation will be particularly

welcome at this time.

My own comments will be largely limited to the notion of the paint system - its concept, its components, its maintenance, and the factors entering into its proper selection by the specifier.

A. THE PAINT SYSTEM CONCEPT

The process of widely choosing a protective coating for a structure to be exposed in a certain environment is an important one in the life of that structure. Like other engineering decisions it requires seasoned judgment and decisiveness in balancing many diverse factors: environment and function on the one hand, versus the broad choice of available protective means on the other. In making these difficult choices, the concept of the integrated paint systems or coating scheme with all of its essential elements is almost mandatory. The experienced specifier realizes that his specification or procurement document must go well beyond the stipulation of a generic or proprietary type of product. In the present state of the art it must deal, not only with the materials used in each coat but also surface preparation, application, thickness and often the steps necessary for proper performance of these materials (Table 1); performance criteria should be used whenever possible.

TABLE 1

A PAINT SYSTEM CONSISTS OF:

A. SURFACE PREPARATION SPECIFICATION
   (Example British 4232/or SSPC-SP)
   (Visual References for Cleanliness)
   (Visual References for Profile)

B. PAINT APPLICATION SPECIFICATION
   (Example - SSPC-PA 1)

C. THICKNESS MEASUREMENT SPECIFICATION
   (Example - SSPC-PA 2)

D. PAINT SPECIFICATIONS
   (Performance or Composition or Proprietary)

Table 2 outlines the elements of one typical paint system, in this case a water-base latex used in comparatively mild inland atmospheric environments. Such a specification occupies only two sides of a single sheet of paper (which can be inserted by reference or in toto in the contract or other procurement document). It includes surface preparation, application, thickness, and sequence of coats.

TABLE 2

SUMMARY OF TYPICAL PAINT SYSTEM

(SSPC-PS XWB1X-79P, Water-Base Paint System)

Section	Requirement	Specification
3.1	Surface preparation	SSPC-SP 6 or SSPC-SP 8
3.2	Pretreatment	Not required
3.3	Paint application	SSPC-PA 1
3.4	Number of coats	3 minimum
3.5	Primer & Intermed.	SSPC-Paint XWB1X Same (tinted)
3.6	Finish coat	SSPC-Paint XWB1X-79P
3.7	Thickness (per SSPC-PA 2)	Primer 2.5 mils; Intermed. 2.0; Finish 1.5; Total 6.0 mils (150µ)
4.0	Inspection	

1. Surface Preparation

We are still learning new things about the practical and theoretical aspects of surface preparation. We do know, however, that surface preparation, together with application and thickness overshadow in importance the usual differences which exist among the kinds of paints used to protect metal. Since removal of the last traces of millscale and other contaminants becomes increasingly difficult and expensive, it is now always economical or prudent to specify the ultimate first quality or white metal blasting cleaning, especially when tolerant paint systems are exposed in milder environments. For this reason both the British Standards and the SSPC specfications recognize lesser degrees of blast cleaning as listed in Table 3.

This table lists the 3 qualities of blast cleaning presented in British Standard 4232, the approximate SSPC equivalents (white metal, near white and commercial as well as a fourth grade - brush off) and the photographic standards which are available as supplemental references.

CORROSION IN CIVIL ENGINEERING

TABLE 3

TYPICAL SURFACE PREPARATION SPECIFICATIONS

CLEANING METHOD	Quality** BS 4232	SSPC*** Spec	Photo Reference SIS 05 59 00 * INITIAL CONDITIONS			
			A Intact Millscale	B Rusting Millscale	C Rusted	D Rusted & Pitted
Solvent cleaning		SP 1				
Hand tool cleaning		SP 2		B St 2	C St 2	D St 2
Power tool cleaning		SP 3		B St 3	C St 3	D St 3
Blast cleaning						
Brush-off		SP 7		B Sa 1	C Sa 1	D Sa 1
Commercial	3rd	SP 6		B Sa 2	C Sa 2	D Sa 2
Near-white	2nd	SP 10	A Sa 2½	B Sa 2½	C Sa 2½	D Sa 2½
White metal	1st	SP 5	A Sa 3	B Sa 3	C Sa 3	D Sa 3
Pickling		SP 8				

* Same as SSPC - Vis 1 "Pictorial Surface Preparation Standards for Painting Steel Surfaces"
** Approximate equivalent in British Standard 4232 "Surface Finish of Blast-Cleaned Steel for Painting"
*** Surface Preparation Specifications of the Steel Structures Painting Council

The development of these particular visual standards represents an excellent example of international cooperation, since most of the pictures were originally developed by the Swedish IVA Corrosion Committee and the remainder, such as the near-white and new commercial blast cleaning photographs, by the Steel Structures Painting Council. Also listed are

TABLE 4

OUTLINE OF A SURFACE PREP. SPEC

SSPC-SP10-67T, Near White Blast Cleaning

1. SCOPE

2. DEFINITION
   End result (photos, if specified)

3. PROCEDURES
   3.1 Sequence
       3.1.1. Solvent clean ⎤ Preliminary
       3.1.2. Power tools   ⎦
       3.1.3. 3.1.3.1 to 3.1.3.7 Methods
              Dry sand, wet sand (sizing)
              Grit, shot (sizing, etc)
              Nozzle - air or vacuum
              Centrifugal recirculating
   3.2 Remove dry abrasive
   3.3 Wet - rinse and inhibit
   3.4 Clean air
   3.5 No damage
   3.6 No rust-back
   3.7 No oil, grease
   3.8 Profile limits
   3.9 Profile measurements
   3.10 Paint promptly

4. SAFETY
   Fire, explosion, dust, filters, goggles, grounding

APPENDIX

A.1 Scope	A.6 Inhibitor
A.2 Where use	A.7 Paint promptly
A.3 Maintenance use	A.8 Photographic references
A.4 Particle sizes	A.9 Other visual standards
A.5 Thickness over peaks	A.10 Alternate % cleanliness

---

the specifications for solvent cleaning, hand tool cleaning and power tool cleaning which are still sometimes used in the United States for rust-tolerant coatings, for mild exposures, for environmental reasons, for preliminaries to blast cleaning, or for maintenance repainting. Table 4 shows the stipulations which experience has shown to be necessary in a typical surface preparation specification. Existing specifications are sometimes used even for new suface preparation equipment such as the proposed KUE air/sand/water method, although they may require supplemental criteria such as the potassium ferrocyanide test for chloride and sulphate ions.

Please note that this typical blast cleaning specification not only includes a definition of cleanliness and the optional use of visual standards, but also specifies preliminary solvent cleaning and power tool cleaning when necessary to remove oil, heavy scale, etc. Profile and protection of the cleaned surface are also covered, and provisions are given for sand, grit and shot, both wet and dry.

2. Specifying Paint Application

In the U.S. some architects, engineers and owners have developed their own paint application specifications. Some prefer to emphasize in the contract a few provisions that pertain to a specific job and then state that in all other respects provisions of a public specification such as SSPC-PA 1 be applied. As shown in Table 5 such a specification provides for application method, storage, mixing, permissible temperature, humidity, cover, damage to existing equipment, striping, contact surfaces, protection of nearby steel during painting, workmanship, inspection, and safety.

3. Specifying and Measuring Film Thickness

The importance of film thickness as a factor in paint life is widely recognized. An interesting relationship is illustrated in Table 6 based on a ten-year report by the SSPC showing approximately 20 additional months of paint life for each mil (25 microns) of paint thickness for these particular paints in these environments.

Measurement of wet film thickness during application can enable the operator and inspector to detect sub-standard thickness at a stage when it can still be corrected. This procedure requires a knowledge of volume percent solids and good technique to establish relationship between wet and dry film thickness.

The specification for measuring dry film thickness, as

TABLE 5

EXAMPLE OF A PAINT APPLICATION SPECIFICATION

SSPC-PA 1   Shop, Field & Maintenance Painting

1. Scope
2. Definitions
3. Procedures
   3.1 Cleaning
   3.2 Pretreatments
   3.3 Storage
   3.4 Mixing
   3.5 Application of paint
      3.5.1 General provisions
         3.5.1.1 Application methods
         3.5.1.2 Temperature
         3.5.1.3 Moisture and humidity
         3.5.1.4 Cover
         3.5.1.5 Damage
         3.5.1.6 Striping
         3.5.1.7 Continuity
         3.5.1.8 Thickness
         3.5.1.9 Recoating
         3.5.1.10 Tinting
         3.5.1.11 Intercoat adhesion
         3.5.1.12 Contact Surfaces
         3.5.1.13 Brush application
         3.5.1.14 Spray (general)
         3.5.1.15 Air spray
         3.5.1.16 Airless spray
         3.5.1.17 Hot spray
         3.5.1.18 Airless hot spray
         3.5.1.19 Roller application
         3.5.1.20 Dip and flow coat
      3.5.2 Shop painting
      3.5.3 Field painting
      3.5.4 Maintenance painting
      3.5.5 Special coatings
   3.6 Drying of painted steel
   3.7 Handling of painted steel
4. Safety precautions
5. Inspection
APPENDIX

TABLE 6

Average paint life vs. thickness (oil and alkyd paints)

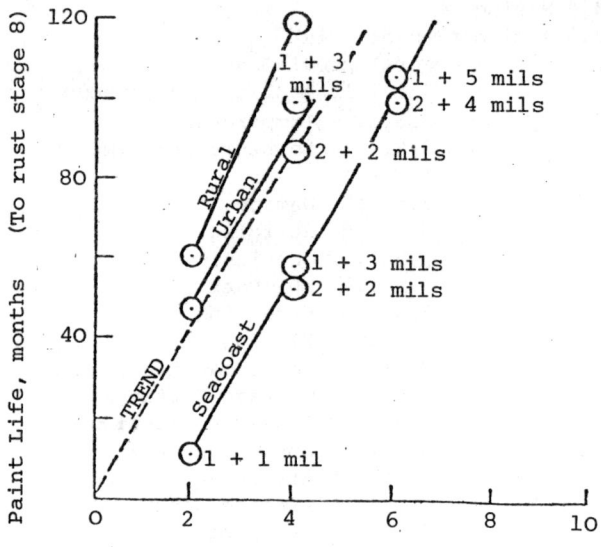

Nominal thickness, mils
Each point represents an average
of 12 panels rated for 10 years.

outlined in Table 7 is quite specific, since some procedures
widely used in the US have led to costly errors, especially
with thin-film protective systems on high-profile surfaces.

TABLE 7

A SPECIFICATION FOR MEASURING

PAINT THICKNESS — SSPC-PA 2

1. SCOPE

2. OPERATING PRINCIPLES

3. PROCEDURES
   3.1 Gage calibration
       3.1.1 Pull-off gages (with plated standards)
       3.1.2 Fixed probe gages (with shim)
   3.2 Thickness measurement
       3.2.1 Type 1 - Pull-off gages
             3.2.1.1 Measurement on substrate (avg.)
             3.2.1.2 Measurement on paint (avg.)
             3.2.1.3 Correct for 3.1.1
             3.2.1.4 Substract 3.2.1.1 from 3.2.1.2
       3.2.2 Type 2 - Fixed probe gages
             3.2.2.1 Measurement on paint film
             3.2.2.2 Average
   3.3 Measurement precautions

4. INSPECTION

5. APPENDIX
   Principles; zero setting; roughness; softness;
   alloys; edge effects; curvature; tilt; residual
   fields; temperature; vibration

4. Paint Procurement by Specification

Procurement of paint in quantities is usually accomplished
by contract. Either composition or performance requirements
are specified or a known proprietary product is named, or a
combination of these stipulations may be used. Since each
type of specification has its limitations and advantages,
there are sharp divergences of opinion regarding which type
assures optimum results. These factors are discussed at
length in the literature and in the Steel Structures Painting
Manual. Table 8 shows the outline of a paint specifcation
now being issued for latex primer for steel surfaces.

TABLE 8

OUTLINE OF TYPICAL PAINT SPECIFICATION

SSPC-Paint XWBIX

LATEX PRIMER FOR STEEL SURFACES

1. SCOPE

2. APPLICABLE DOCUMENTS

3. GENERAL REQUIREMENTS

4. DETAILED REQUIREMENTS
   4.1 Materials
   4.2 Composition requirements
        % Non-volatile, % pigment,
        Pigment-volume conc.,
        % Solids, etc
   4.3 Preservative
   4.4 Color
   4.5 Condition in container
   4.6 Application properties
   4.7 Freeze-thaw
   4.8 Wet adhesion
   4.9 Package stability
   4.10 Mechanical stability
   4.11 Blister resistance
   4.12 Coalescense test
   4.13 Mud-cracking
   4.14 Salt-fog test
   4.15 Flash-rust resistance test
   4.16 Wet abrasion resistance
   4.17 Fineness of grind
   4.18 Drying time
   4.19 Flexibility
   4.20 Mildew resistance
   4.21 Consistency
   4.22 Specular gloss

5. EXTERIOR EXPOSURE TEST
   5.1 Bid sample
   5.2 Applicators

6. MARKING OF CONTAINER

7. NOTES
   Intended use; unit of purchase; waiver; storage;
   ordering data; temperature

(There is much sentiment in regulatory circles in the U.S. for eliminating all volatile organic compounds, resulting in intensive attempts to develop suitable water-base coatings for steel.)

It should be noted that most of the requirements of this specification deal with performance requirements such as adhesion, stability, blister coalescense, mud cracking, salt fog tests, flash resistance, etc. Some general composition requirements are also necessary in most specifications, however, in the present state of the art to assure proper percent solids, etc.

With other generic types of coatings other appropriate requirements must be added, such as specifying that zinc-rich paints must contain some zinc and that some epoxy resin be present in the specified epoxy paint. (This paper is not completely consistent in its effort to apply the term paint to the wet product and confine the use of the word coating to the applied film.)

Table 9 gives an outline of the 18 generic types of SSPC paint systems. Please note that there may be from 1 to 6 paint systems within any one generic type and that each paint system may cite from 1 to 4 individual paint specifications. This diversity provides the purchaser with a wide choice of systems to meet his particular needs.

In general each specification pertaining to a particular generic type of paint or paint system issued by SSPC is originated and maintained by a consensus committee of specialists concerned with that particular kind of coating. Thus there are SSPC advisory committees for zinc rich, epoxies, chlorinated rubbers, silicone alkyds, water base, urethanes, etc. each having a balanced representation among raw material suppliers, paint manufacturers, applicators, the public interest, and users, both governmental and industrial. Typically such a committee would be made up of 1 to 2 dozen specialists who develop case histories, guides, paint specifications, paint systems and recommendations for the SSPC field evaluation program. All documents emanating from such committees first undergo a consensus process within the committee before the guide or specification is submitted to the research and executive committees of the SSPC for final approvals. You can well imagine that, although this system is a very thorough one, it can often be time consuming.

TABLE 9

OUTLINE OF SSPC PAINT SYSTEMS

No.	SSPC SYSTEMS	GENERIC	USES
1	PS 1.00–1.06	Oil base	For weather-exposed wire-brushed steel
2	PS 2.00–2.05	Alkyd	For weather exposure
3	PS 3.00	Phenolic	For high humidity, condensation or fresh water immersion
4	PS 4.00–4.05	Vinyl	For immersion, industrial or chemical exposure
6	PS 6.00–6.03	For vessels	Bottoms, boottopping, topside, super-structures
7	PS 7.00–7.01	Shop paints	For non-corrosive interior or short-term exterior
8	PS 8.01	Rust Prev. compounds	For temporary protection or sheltered locations
9	PS 9.01	Asphalt	For weather and corrosive atmospheres

10	PS 10.00-10.01	Coal tar	For underwater, underground or corrosive environments
11	PS 11.01	Coal tar epoxy	For fresh or salt water, chemicals, underground
12	PS 12.00-12.01	Zinc-rich	For marine, abrasion, immersion, chemicals
13	PS 13.00-13.01	Epoxy	For water immersion, chemical, industrial, or marine use
14	PS 14.01	Alkyd or asphalt	For steel joints, for interior use
15	PS 15.01	Chlorinated rubber	For chemical, marine, moisture
16	PS 16.01	Silicone-alkyd	For gloss and color retention, chalk resistance
17	PS 17.01	Water-base	For air pollution control areas; weather-exposed
18	PS 18.00	Urethane	For weathering, chemical resistance, low temperature curing, high-build

CORROSION IN CIVIL ENGINEERING

B. CHOICE OF A PAINT SYSTEM

Because of the wide diversity of available systems, the
adverse conditions under which they must sometimes be applied, environmental restrictions, and the need for mini-
mizing maintenance, the choice of a suitable paint system is
not always an easy one. There is, of course, no one "best"
paint system but rather a dynamic competition among alternative materials and methods whose choice often depends both
upon technological factors and policy considerations.

1. Environment

Among the purely technological factors, environment is
usually the controlling one. Table 10, for example, illustrates one rule-of-thumb which advocates that steel need not
be painted at all when corrosion rate is uniform and below a
certain level; when environment is too severe, on the other
hand, alternate materials of construction should be considered instead of painting. Fortunately improvement in
coatings technology are extending both the upper and lower
range of economic coatings performance.

Table 11 illustrates the wide range of atmospheric corrosiveness as measured by the American Society for Testing
Materials several years ago, varying a thousand-fold in
various parts of the world. On the one hand increased atmospheric contamination has resulted in acid rain waters, while

TABLE 10

CORROSION RATE VS. SUGGESTED PROTECTION

Corrosion rate per year		Suggestion
Uniform	Pitting	
(125μ)<5 mils	None	Paint only for appearance
5-10 mils (125-250μ)	Slight	Paint with economical systems
5-10 mils (125-250μ)	Up to 50 mils (1250μ)	High performance paint systems
>10 mils (250μ)	>50 mils (1250μ)	Consider alternate materials

TABLE 11

SOME MEASURED CORROSION RATES (a)

Site No.	Location	Type of atmosphere	Relative steel	Rating zinc
1	Normal Wells, N.W.T.	Rural	0.02	0.2
2	Saskatoon, Sask	Rural	0.2	0.2
9	State College, PA (b)	Rural	1.0	1.0
17	Pittsburgh, PA (roof)	Industrial	1.8	1.5
18	London (Battersea)	Industrial	2.0	1.2
27	Bayonne, New Jersey	Industrial	3.4	3.1
28	Kure Beach, NC (800 ft site)	Marine	3.6	1.9
31	London (Stratford)	Industrial	6.5	4.8
33	Point Reyes, California	Marine	9.5	1.0
37	Kure Beach, NC (80 ft site)	Marine	33.0	6.4

(a) Adapted from ASTM Materials Research & Standards, December 1961, page 977.

(b) State College, PA, taken as unity. Fortuitously, a corrosiveness of 1.0 represented about 1 mil loss the first year.

on the other many industrial locations are less contaminated than in the recent past.

Table 12 illustrates the "zone defense" concept of the SSPC. It suggests a preferred system along with some alternative systems, for consideration in rural, urban, commercial and marine environments and in several special exposures. A more detailed version of this type of recommendation is given in Volume 2 of the Steel Structures Painting Council manual. Another type of detailed recommendation is illustrated schematically in Table 13 giving alternate recommended systems for 500 types of surfaces, listed alphabetically in each of 10 types of exposures.

TABLE 12

SOME TYPICAL RECOMMENDATIONS FOR COATING STRUCTURAL STEEL

Zone *	Environment	Preferred system	Alternates
1A	Interior, normally dry (or temporary protection) Unusual in hwy. work, very mild (oil base paints would last 10 yr or more)	One coat of fast-drying shop paint (example: SSPC-Paint 13) over nominally hand-cleaned steel. Finish coat optional (see SSPC-PS 7.01)	(1) Other one-coat primers (example: TT-P-636) (2) Rust proofing (SSPC-PS 8.01), or (3) More durable systems as per Zone 1B, or (4) Approved proprietary paint
1B	Exteriors, normally dry (includes most highway areas where oil base paints now last 6 yr or more)	Apply 2 coats of oil base primer (example: SSPC-Paint 14) over wire-brushed steel. 1-2 finish coats of long oil alkyd (SSPC-Paint 101 aluminum or SSPC-Paint 104 white, gray or green) 4.0 mils or more thickness (5.0 mils for 4 coats). (See SSPC-PS 1.01, 1.02 or 1.03)	(1) Blast clean (SSPC-SP 6) and use same paints or shorter oil alkyds. (2) Alternate primers (SSPC-Paint 2; TT-P-57, Type I; AASHO M72-57, Type I or II; or TT-P-615, Type V) or (3) Alternate intermediate TT-P-86, Type II or non-leafing aluminum, or (4) Equivalent state system, or (5) Same systems as Zone 2A or 2B, or (6) Proven proprietary system.

46

| 2A | Frequently wet by fresh water Involves condensation, splash, spray, or frequent immersion. (Oil base paints now last 5 yr or less) | Near-white blast clean surface; 4 coats (4.5 mils) of vinyl system (example: SSPC-Paints 8 or 9) (See SSPC-PS 4.04 or 4.02) | (1) Pickle (SSPC-SP 8) instead of blast cleaning.
(2) Alternate vinyls are VR 3 or approved proprietaries.
(3) Epoxy system guide (SSPC-PS 13.00) coal tar epoxy (SSPC-PS 11.01), chlorinated rubber system, or approved proprietary system. |
| 2B | Frequently wet by salt water Involves condensation, splash, spray or frequent immersion. (Oil base paints now last 3 yr or less) | Near-white blast clean surface; apply zinc-rich primer (example: SSPC-PS 12.00 or MIL-P-23236 or California Highway Spec. 66-G-55) followed by approved wash primer and finish coat. (Example: SSPC-PT 3 plus SSPC-Vinyl Paint 8 or 9, 3+ mils) Assure satisfactory adhesion of finish coats. | (1) Use finish coats with same vehicle as zinc-rich primer (inorganic, epoxy, chlorinated rubber, vinyl, etc)
(2) Use vinyl paint system with wash coat and inhibitive primer (example: SSPC-PS 4.01 or 4.03)
(3) Use as alternate finish coats or by themselves, coal tar epoxy (SSPC-PS 11.01), epoxy (guide SSPC-PS 13.00), or approved chlorinated rubber system, or other proven proprietary system. |

Continued on next page

TABLE 12 CONTINUED

Zone *	Environment	Preferred system	Alternates
3	Chemical exposures (Acidic, alkaline, oxidizing, solvents, etc)	Same as for Zone 2B, but with chemically resistant finish coat system specially chosen to protect primer and base metal against specific chemical agent. (Zinc-rich unsatisfactory for very acid or very alkaline conditions.) Assure satisfactory adhesion of finish coats.	Same choices as for Zone 2B, but with special finish coats. (1) Coal tar epoxy (SSPC-PS 11.01) (at least 16 mils). (2) Straight vinyls for acid and alkali (SSPC-PS 4.01) or 4.03). (3) Epoxies for alkalies, salts, aliphatics, acid splash; not for strong solvents. (4) Neoprenes and other proven proprietary systems to resist specific conditions.
4	SPECIAL CONDITIONS Painting galvanized steel	Solvent clean to remove oil and grease. Wire brush to remove any rust. Apply zinc dust-zinc oxide paint TT-P-641 (Type II for new steel, Type I for old, as per SSPC-PS 2.05 and 1.04). Somewhat better adhesion if surface is weathered before painting.	(1) Chemical pretreatment of new work by commercial hot phosphate or wash primer. (2) Zinc-rich primer (example: Guide SSPC-PS 12.00). (3) Prime with SSPC-Paint 5. (4) Prime with proven proprietary cement-base, polyvinyl acetate emulsion, or acrylic latex.

Mildew	After surface preparation, wash mildewed surface with trisodium phosphate and dry. Add mildewcide to each coat of paint (example: 8-quinolinoleate). Vinyl, chlorinated rubber resins, and barium metaborate and zinc-rich pigmentations tend to resist mildew.	Alternate mildewcides and fungi-cides include copper naphthenate, chlorinated phenols, phenyl mer-curic dodecylsueinate, proprie-tary agents. Add in amount recom-mended by the manufacturer.
Temporary protec-tion and rust-proofing	See system for Zone 1A. Also see SSPC-PS 8.01, "Rust Preventive Compounds" (thick non-hardening films over minimum surface sur-face preparation)	Soft, heavy or hard film compounds as per 52-MA-602, Type B, C, or D; or use proprietary rust-proofing compounds.
Painting welds	Before welding, do not paint within 2 in. of edges. Blast clean after welding. See SSPC-PA 1, Sections 3.5.2.4 and 3.5.2.5.	Chip and wire brush weld thoroughly. Wash with 5% phosphoric acid and rinse. See SSPC-SP 1, Section 3.1.6.

* These are intended as specific exposure zones of the portion of the structure under considera-tion rather than geographic zones. Severity of exposure can change sharply over very short dis-tances due to such factors as wind, spray, condensation, and use of de-icing chemicals.

Such guides and specifications are intended to aid the specifier in selecting a paint system (including surface preparation, application and coatings) without having to reproduce all of the detailed provisions of those specifications. Such guides, of course, are no substitute for the knowledge and judgment entailed in an intelligent choice.

2. Other Factors in System Selection

Although environment is usually the primary factor in the choice of a coating system, other factors such as costs, appearance, design, available facilities and availability of specifications must also be considered.

Costs per square foot per year should ideally be minimized over the projected life of the structure. A number of the more sophisticated engineering methods* are sometimes justified, since the cost of coatings during the total life cycle cost of a structure often exceeds the initial cost of the structure itself. In the U.S. these cost calculations are currently simplified by the circumstance, perhaps fortuitous, that money interest rates are of the same order of magnitude as the inflation rate, so that a dollar set aside today at current interest rates may be presumed to have the same purchasing power when it is ready to be spent at some future date.

One of the simplest means for comparing costs of two or more alternative coating systems is illustrated in Table 14. This hypothetical example illustrates how, in a certain Zone 2 type of environment it is more economical to use a "deluxe" paint system (for example top-coated zinc rich over near-white blast cleaned steel as per SSPC-PS 12.00) than a "low costs" system (such as SSPC-PS 7.01) in a typical case where the life of the latter is five years or less. In many U.S. environments, however, a coating life of 7 or more years can be obtained with a simple alkyd paint system.

Here it becomes more difficult to justify on a purely economic basis the high initial cost entailed in a potentially more durable system. Such systems usually involve more expensive surface preparation and application and are less tolerant of lapses in workmanship. The cost of the paint itself is, of course, a relatively minor consideration since surface preparation and application costs are usually from 2 to 10 times the material costs.

* Life cycle costing; cost/benefits ratio; discounted cash/ flow; capitalized cost method; payback time; calculated risk; net present value of alternative cost; equivalent uniform annual cost; or return on incremental investment.

TABLE 13

INDEXED PAINTING RECOMMENDATIONS FOR STEEL

(From SSPC Manual – Volume 2, Table III)

SURFACES (500 items)	TEN EXPOSURES					
	Rural	Ind.	Water		High Humidity	5 Other zones
			Fresh	Salt		
Abattior, exterior	PS 1, 2	PS 1, 2			PS 3, 11	
Abraded surfaces	See Table II – Abrasion resistance					
Acid barge hull, exterior			PS 3, 4, 9, 11	PS 4, 9, 11		
interior	See Corrosive chemical conditions					
Aeration equipment					PS 3, 4, 9, 11, 12	
Air duct exterior, black iron, etc	PS 1, 2	PS 1, 2			PS 3, 4, 9, 11, 12	
(500 other items)						

TABLE 14

TYPICAL

COATING COSTS

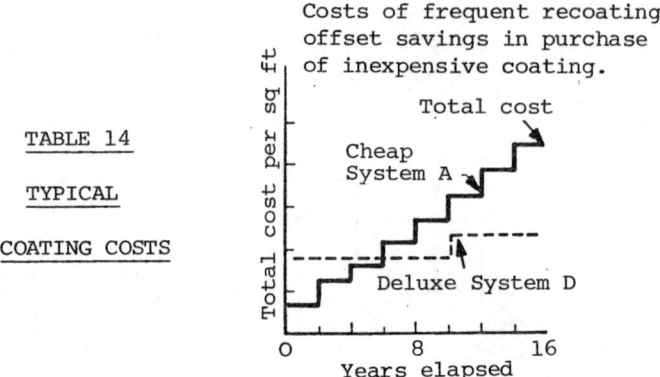

Costs of frequent recoating
offset savings in purchase
of inexpensive coating.

Total cost

Cheap System A

Deluxe System D

Total cost per sq ft

O   8   16

Years elapsed

TABLE 15

SOME DESIGN FEATURES TO FACILITATE PROTECTION

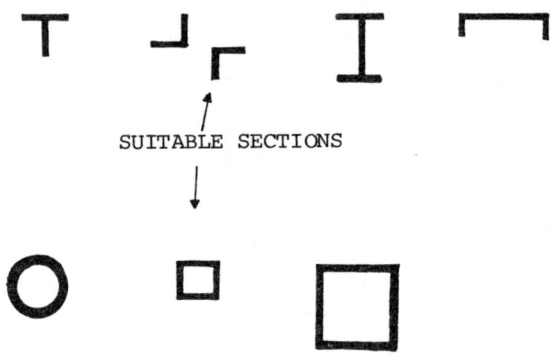

SUITABLE SECTIONS

TUBES AND CONTINUOUSLY WELDED BOX
SECTIONS SHOULD BE SEALED AT ENDS

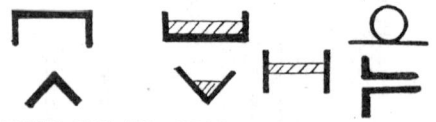

CORRECT WAY TO   WRONG WAY TO USE CHANNELS,
USE CHANNELS   ANGLES, AND TUBES.
AND ANGLES.

Appearance is a major consideration in painting many
bridges, tanks, refineries, plants, etc. and color is often
chosen to harmonize or contrast with adjacent topographic or
community features. Appearance plays an increasingly important role in structural painting, but corrosion prevention
is necessary to retain that appearance.

Engineering design of a steel structure can often render
protection by paint either relatively straightforward or
almost impossible, as pointed out by the British Iron and
Steel Research Association. Table 15 illustrates some of
the many features now being adopted in the U.S. to facilitate protection and maintenance of structures. Like BS 5493,
the SSPC manual discusses alternative coating practices
which may actually influence the design of structures whose
surfaces are below ground, faying, or exposed to salt drip-
pings, condensation, debris, etc.

Facilities available to local painting contractors and
fabricators sometimes limit the type of surface preparation,
application, or coatings which can be specified in some
parts of the U.S.

Availability of specifications is essential for many
types of public and private structures in the U.S., where
policy encourages the procurement of materials and services
on an open competitive basis. For this reason the SSPC, as
well as the U.S. Government is continually involved in
improvement of specifications, particularly for surface preparation and for newer coating systems. Performance criteria are used whenever possible, but in practice composition requirements must also be included. Some users also
prefer to specify by proprietary product name based upon
proven past performance or by qualified products lists.

C. "TRADE-OFFS" IN MAINTENANCE PAINTING

In maintenance painting as well as in engineering of new
structures, the personnel involved in corrosion protection
and protective coating must work hand in hand with management in recognizing the "trade-offs" inherent in painting
policy. Some of these considerations are outlined in Table
16.

Environmental constraints for example are causing the
SSPC and others in the U.S. to look carefully at alternatives to sand blasting, at new inhibitive types of pigment
and at new coatings with very low volatile organic compound
content. Much of our present surface preparation work, for
example, is concerned with the substitution of low-free-
silica abrasives for sand, with the use of metallic and

CORROSION IN CIVIL ENGINEERING

TABLE 16

SOME "TRADE-OFFS" IN MAINTENANCE PAINTING

1.	COATINGS DURABILITY	VS.	ENVIRONMENTAL RESTRAINTS
2.	INITIAL COST	VS.	MAINTENANCE COSTS
3.	SHOP PAINTING	VS.	FIELD PTG. VS. DEFERRED PTG.
4.	PERFORMANCE SPECS	VS.	COMPOSITION VS. PROPRIETARY
5.	CONTRACT	VS.	"IN-HOUSE" PAINTING
6.	DESIGN	VS.	COATING EFFECTIVENESS
7.	APPEARANCE	VS.	PROTECTION
8.	QUALITY	VS.	COST OF PERSONNEL COMMITMENT

other re-cyclable abrasives, with new wet blast techniques such as the British KUE water/air/sand method, and with entirely new surface preparation concepts.

A special project PACE (Performance of Alternate Coatings in the Environment) has been continued by SSPC in order to compare conventional time-honored coating pigmentation versus non-lead, non-chromate pigmentations on which less experience is available, such as molybdates, phosphates, metaborates, calcium boron silicate, calcium phospho silicate zinc sulpho oxide complex, etc. This program is also looking at alternative coatings tolerant of non-sand-blasted surfaces.

In anticipation of possible further air pollution legislation the SSPC and others are looking more carefully at the compromises involved in use of water-borne coatings which admittedly have, at the present time, shorter coating life than those to which we are accustomed. Even oil-base coatings and long-oil alkyd modifications are being considered because of their tolerance for non-sand-blasted surface and low organic volatile content.

Other "trade-offs" in Table 16 which are the subject of continual dialogue and re-appraisal in the U.S. include:
• The choice of an initial paint system should optimize the present value of the total initial protection plus maintenance costs for the life of the structure.
• Shop painting versus field painting versus deferred painting. In new work it is common to carry out the surface preparation and priming in the shop, touch-up in the field, and finish coating after erection. Other practices are

sometimes followed, such as deferring the second coat of
paint for several years when the original primer was a zinc
rich. This practice may have a tax advantage and has often
proven technically feasible.

● Performance specifications versus composition versus
proprietary are discussed herein and at greater length in
the SSPC manual. The U.S. Government and the SSPC attempt
to combine the advantages of these three methods, by using
proprietaries which meet the performance requirements and
yet stay within the required broad compositional limits.
"Pilot" formulations are sometimes specified for use as control comparisons in accelerated salt fog tests.

● Contract painting versus "in-house" painting. Many large
users have their own highly qualified full-time crews whereas others favor the use of a knowledgeable contractor whose
sole business is painting. The painting contract usually
provides for either a "cost plus" basis (at a competitive
overhead rate) or a "hard dollar" basis, usually based upon
competitive bidding.

● Quality painting is usually the result of a commitment by
management for the selection, training, certification, and
support of properly qualified personnel, including supervision, engineering and inspection. If these capabilities
are not present "in-house" then they too can be obtained
through judicious contract for new or maintenance painting.

D. FURTHER RESEARCH NEEDED

To serve the needs of the specifier and the ultimate user,
it is recommended that guides and specifications be further
developed and more widely used. In addition, however, further research is needed in order to fill in the many gaps in
our knowledge and creativity applied to coatings technology.

In some related areas of technology an impressive backlog
of knowledge has already been developed by academe, industry
and government in metallurgy of the substrate, fundamental
corrosion mechanisms, development of new polymers, etc.
Surely the specifier will be the beneficiary of this work.
In the meantime, however, there is a more immediate need for
applied research which will tell us more about the three
types of interface in Table 17.

Studies of the interface between the coating and its
steel substrate by the Paint Research Association, the SSPC
and others are just beginning to yield information on profile, effect of ionic contamination and new methods of surface preparation. These give promise of providing the
specifier with much more reliable means and criteria than

CORROSION IN CIVIL ENGINEERING

TABLE 17

RESEARCH NEEDED AT COATINGS INTERFACE

---

1. Interface between coating and substrate
2. Interface between coatings
3. Interface between coating and environment

---

are now available for preparing the surface for coatings application.

Applied research on the interface between paint and paint is exemplified by the extensive studies being conducted on the topcoating of zinc-rich paints with the available generic types of finish coats. Further work should also be done on the mechanisms involved in maintenance repainting.

Creative and imaginative research on the interface between the coating and the environment is necessary on the one hand, to protect vulnerable materials of construction, like steel, from an increasingly hostile environment. To aid the specifier, however, the reciprocal effect of the coating upon the environment must also continue to be the subject of intensive applied research: to develop silica-free surface preparation methods; to devise durable low-solvent coatings; to provide products which are non-polluting, ecologically acceptable, non-toxic, and formulated from abundant, low-energy-consumption types of raw materials; and to explore innovative approaches to the problems of materials protection.

E.   ACKNOWLEDGEMENTS

I do hope that my remarks have not served to illustrate too vividly the famous remark by your George Bernard Shaw that we are two great countries separated by a common language.

We have cited the specifications and guides of the Steel Structures Painting Council, which represents a broad consensus of most of the major organizations in the U.S. concerned with the use of coatings to protect steel. These organizations include the major associations representing the steel industry, petroleum, water, storage, aluminum, zinc, lead, paint industry management, paint technology, paint applicators, corrosion engineers, materials testing, raw material suppliers, ship builders, power plants, highways, turnpikes, and ten agencies of the U.S. Government. We especially wish to congratulate the British organizations who have done such excellent work in our field of common interest, particularly to Mr Chandler and his committee on BS5493, the Paint Research Association, to the Institution of Corrosion Science and Technology, and the Institution of Civil Engineers.

5 DIN standard in respect of corrosion in civil engineering

DIN 55 928: Protection of steel structures from corrosion by organic and metallic coatings

R. J. HECKER, Spies, Hecker GmbH

SYNOPSIS. The German standard on the protection of steel structures from corrosion has recently been completely revised and made more comprehensive: the new version was published in November 1978. The new standard covers the whole corrosion protection process from the design requirements, through system selection and surface preparation, to the application and testing of coatings. Reference is made to existing standards and codes in matters such as rust and cleanness grades (the Swedish pictorial standard is used), metal protective coatings and test methods, but many other standards and rules have been revised and incorporated in the new standard itself. The standard is inevitably now very bulky but tables are used throughout to help the reader to use the information.

Under this heading, there has been existing a standard in Germany since 1959 governing the execution of corrosive protection work in the sector of steel structures among the parties involved.

In the course of time, the requirements for corrosive protection have been increased and the importance of the appropriate preconditions have been recognised in respect of an economical and durable corrosive protection.

DIN 55 928 was therefore completely revised and supplemented. This measure, at the same time, aimed at standardising the multiplicity of existing rules and regulations of the most different types of interest groups.

The present work, having been concluded and published in two sections, was performed by the study group for standardisation 10 "Protection of steel structures from corrosion"

CORROSION IN CIVIL ENGINEERING

under the supervision of Federal Railway Director Landwehr
and in co-operation with about 100 members from the fields
of:
Steel construction;
Paint varnish, lacquer and raw material manu-
facture;
Painting contracting;
Galvanising;
Material testing.

The complete new version of Standard 55 928 is scheduled to
be published in November 1978. The new DIN Standard 55 928
comprises 8 sections covering the following problem areas:

Section 1 explains the purpose and application of the standard and covers the terms and definitions.

Section 2 deals with the corrosion-proof design of steel
structures.

Section 3 describes the planning of the corrosion protection
work.

Section 4 describes the preparation and testing of surfaces.

Section 5 deals with coating materials and protective systems.

Section 6 describes the execution and control of the corrosion protection work.

Section 7 contains the technical regulations for control surfaces.

Section 8 deals with the corrosion protection of load-bearing
thin-walled components.

I would now like to discuss these sections of the standard
somewhat more in detail.

The definitions laid down in section 1 cover the application
of the present standard to steel structures requiring a
static calculation or a building licence; the concept of corrosion protection; coatings and the various types of corrosive hazards, including special hazards of a chemical,
mechanical or thermal nature.

Section 2 features basic principles of corrosion-proof
design which the design engineer should take into account.
The design of structural components with due consideration
for the requirements for protection from the corrosive
environment at the place of construction as well as the need
for accessibility to protective treatment exert a consider-

able influence on the service life and thus the economy
and efficiency of the structure.

Sketches from practical structures show details which avoid
corrosion protection problems by appropriate design. The
engineer should make correct use of these examples in his
own designs.

If it should be impossible to put the design suggestions
into practice in a particular case, the structural components involved will have to be provided with additional
corrosion protection.

Section 3 under the heading "Planning of the corrosion protection work" explains the factors to be taken into consideration in the planning stage. It describes the compilation
of planning parameters, the consequent specification of protective systems, including the preparation of the surfaces
concerned, and the time schedule for the respective operations.

Schematic phase sequences for the first protective measures
and for the relevant maintenance work are illustrated pic-
torially.

A particularly important part of section 4 "Preparation and
testing of surfaces" is the required degree of cleanness.
This depends on the following factors:

    a) the type of corrosion protection system selected
    b) the operating conditions to be expected
    c) the possible methods of removing rust
    d) the initial condition of the surfaces to be treated.

At this point, the various methods of removing rust are described in relation to the degrees of cleanness achievable
with them.

Tables are used to illustrate the respective requirements.

For the purpose of assessing the surfaces before treatment,
this standard is based on the Swedish SIS 055900 "Degrees of
rusting of steel surfaces and quality grades for the preparation of steel surfaces to be treated with corrosive protection paint coatings".

A table allows other degrees of derusting or other quality
grades to be compared.

Section 4 of DIN 55 928 replaces the standard DIN 8202
"Testing the cleanness and roughness of blast-cleaned metal
surfaces".

CORROSION IN CIVIL ENGINEERING

This section 4 of DIN 55 928 includes a supplementary sheet with photographic samples for comparison of the rust degrees A, B, C, D with the original condition of the uncoated steel surfaces as well as standard degrees of cleanness according to DIN 55 928, section 4. These photographs are identical with the corresponding photographs of the Swedish standard SIS 05 5900.

As regards section 5, I would like to furnish you with a brief summary of the contents:

Section 5 covers coating materials (the relevant definitions are specified in DIN 55 945) and protection systems for efficient and economical protection of steel structures from corrosion.

The selection of these materials and systems is made:

1. as a function of the corrosive hazards to be expected
2. by taking into consideration the service life of the component.
3. by estimating future maintenance costs.

High-quality protective systems usually increase the costs a little, but considerably prolong the period of protection.

Breakdown of the Standard

1. **Purpose and Application**

   Specification and application of coating materials and protection systems for efficient and economical protection from corrosion as a function of the hazards to be expected.

2. **Jointly applicable Standards**

   About 50 German standards are indicated which are jointly applicable. I shall mention some important ones of them a little later.

3. **Coating Materials and Processes**

3.1 Materials and processes or methods for temporary protection: shop primer, wash primer, peel-off protective lacquer and the like.

3.2 **Coating Materials for Primer Coatings and Top or Finish Coatings as well as for the Protection of Edges**

   These terms are defined.

3.2.2. Pigments for Primer Coatings

Only the common corrosion protection pigments are indicated.

3.2.3. Pigments and Fillers for Top or Finish Coatings

As regards such top coatings, the selection of the pigments is determined by the corrosive hazards, by the vehicle or binding agent and the colour desired. A table shows the customary pigments and fillers.

3.2.4. Binding Agents for Primer Coatings and Top Coatings

Here, there are indicated the customary vehicles characterised briefly by their suitability for protection against corrosion. A table features guideline values for the coating thickness, drying time and temperature tolerance. It also shows the compatibility of vehicles with the corrosion protection pigments.

4. Metal Coatings and Substances for covering Metal Coatings

Tables show methods (e.g. hot dip coating, hot galvanising), quality grades, covering or coating thickness, and general instructions.

The "duplex method" is recommended and data on suitable and appropriate vehicle systems are indicated.

5. Corrosive Protection Systems

Successive coatings are to be applied with minimum possible delay.

5.1 Coating Thickness

Among other factors, the protective effect and efficiency decisively depend on the thickness of the coating.

The term "nominal coating thickness" has been newly introduced here. The former term was "minimum coating thickness". The nominal coating thickness is the coating thickness of the system or single coating layer which provides sufficient and economical protection against the corrosion hazards to be expected. It is regarded as being complied with if 10% of the measuring values at the most fall below the nominal value by 10% at the most. The coating thickness is controlled by non-destructive measurement. The

following number of measurements is recommended:
20 measurements per 100 m^2 of measuring area in connection with objects exceeding an area of 5,000 m^2.
Smaller objects do not require such a quantity of
measuring points.

5.2   Corrosion Protection of hollow cases and boxes,
        hollow components and special structural components

Coating systems are recommended, depending on the
designs, structures and hazards involved. This also
applies to structural components within buildings, to
steel structures with components of concrete, and to
steel structures with fire protection coatings.
Special protective systems are indicated for steel
hydraulic structures.

6   Testing of Coating Materials and of Coatings

Tests are necessary in the following cases:

a) If insufficient practical experience is available
b) For the quality control of current deliveries
C) Identity tests for comparison with former
   deliveries

The test results may be compared only if they have
been ascertained on the basis of the same test
methods.

Explanations as to Section 5 of DIN 55 928

As various and different official German rules and regulations, conditions of delivery and works standards of major
purchasers were to be compiled into one binding standard,
the particular requirements of the relevant parties involved
had to be taken into consideration, too.

The description of substances and materials as well as the
multiplicity of schematic representations in the form of
tables are intended to help in the preparation of tenders.

Due to these requirements, it could not be avoided that this
standard has become unusually extensive and that it somewhat
features the character of a textbook. Subject extent, however, is justifiable because quite a number of official and
industrial conditions of delivery will be replaced by the
revised standard 55 928 as soon as it comes into force.

Section 6 contains specifications for the execution and control of the corrosion protection work. Among other items,
this asks for the qualification of the executing personnel

both from the technical and the personal point of view as a precondition for the performance of such work. As regards work to be performed in accordance with TRBF 401 and 402 "Interior coating of tanks and containers for the storage of inflammable fluids", one even goes a step further here in Germany. The applicator has to have an approval by the Technical Supervision Authorities which requires proof of the knowledge of all corresponding rules and regulations, of the practical qualifications as well as of the relevant processing equipment.

Concerning the application of fire protection coatings, which which also comprise the protection from corrosion in the case of steel, the manufacturers of such materials are obliged in accordance with the construction supervision regulations to instruct the contractors in the correct application of the coating materials and their testing on the component. In this connection, proof is to be furnished for every group of contractors.

Section 6 furthermore contains a description of the various methods of application as well as instructions on the supervision and testing of the execution of corrosive protection. In this connection, reference is made to a standard "DIN 53 151" for testing the adhesion of paint coating systems which is contested in practice. It deals with the cross-cut test on the component. The DIN standard is based on laboratory tests. There are no indications of time-related data as to the implementation of the test.

Among other particulars, the explanations of this DIN standard read as follows: "In addition to the adhesion of the paint coating on the ground, the ductility, hardness and strength or stability will be incorporated in the test result to a certain extent. As is known by experience, for example, ductile paint coatings with a low degree of adhesion do not burst off so much on the occasion of a cross-cut test as very brittle paint coatings with a good adhesion". This fact alone shows the problematic nature of this subject.

A deficiency developing on the component within the period of guarantee may become a point of controversy. In order to be able to judge the cause of such a deficiency as objectively as possible, technical rules for control surfaces have been established in section 7 of DIN 55 928.

As coatings applied by third parties are increasingly going to be overpainted, with the respective guarantee being assumed, the system of the double-controlling surfaces has been newly incorporated. Here the standard reads: "Concerning the coating of surfaces, having previously been provided with finish and/or primer and top coatings by third parties, double-control surfaces must be provided as follows:

Control Surface A	The specified further coating or over-painting is applied over the existing partial coating following its preparation according to the contract.
Control Surface B	Upon removal of the existing partial coating and after having established the specified degree of cleanness of the steel surface, the complete corrosion protection system is to be applied so that there will first be applied a partial coating of the same type as the removed coating."

Reports are to be made about these control surfaces during every processing sequence. These reports are to include all important data and information both of technical and of environmental nature, such as temperature, air humidity, etc., which are to be confirmed by the authorised personnel of the contractor as well as by the authorised personnel of the supplier of the paint coating materials.

An instruction is given for the evaluation of these controlling planes which also takes into consideration the possibility of damage by third parties, e.g. changes in environmental influences.

Section 8 of DIN 55 928 specifies the corrosion protection by coatings and coverings for thin-walled loadbearing structural steel components mainly stressed by static loads. The wall thicknesses are under 3 mm.

Reference is largely made to the sections of DIN 55 928 already mentioned.

Depending on the risk of corrodibility distinctions are made between various cases of corrosion hazards as described in section 1, para 3.

The resulting corrosion protection classes call for corresponding protective systems. They are intended to form part of design documents to be submitted for obtaining the building permit.

In the same way as in section 5, tabular compilation helps in the determination and selection of the appropriate corrosion protection systems.

Here, as in sections 4 and 5, the jointly applicable standards are: DIN 8565 "Protection of Steel Structures from

Corrosion by means of a Thermal Spraying of Zinc and Aluminium"; DIN 17 162 - section 1 - "Flat-steel Products; hot-
galvanised Strip and Sheet Metal of soft, unalloyed Steels"
Technical Conditions of Delivery and DIN 50 976
> Corrosion protection: Requirements for Zinc Coatings
> on Objects of Iron being hot-galvanised as Finished
> Products.

Relevant standards to be mentioned in connection with DIN 55
928, in the first instance, are testing standards. In addition to DIN 53 151 "Cross-cut Test on Paint Coatings and
Similar Coatings" already mentioned before, a further test
of the adhesion has been established in DIN 53 232 "Measurement of the Adhesion of Paint Coatings and Similar Coatings
by the Pull-off Method".

DIN 50 981
> 50 982 section 2 and
> 50 985

cover the measurement of coating thicknesses on various
types of surfaces such as iron and non-ferrous metals.

DIN 50 018 has been introduced as standard for corrosion
tests. Exposure in a damp heat alternating atmosphere with
the addition of sulphur dioxide at specified temperature
cycles is known as Kesternich Test.

DIN 53 156 describes the "Cupping Test according to
Erichson on Paint Coatings with Optical Evaluation".

DIN 53 159 and DIN 53 223 specify the chalking of paint
coatings on the basis of different methods.

DIN 8201 deals with methods of abrasive blasting; this standard plays an important role in DIN 55 928. It is to be
taken into consideration in connection with the Ordinance
on Hazardous Operating Media - appendix II, No. 3, para 3.3,
Prohibition of using silicogene-type abrasive blasting
agents.

This is the end of a brief survey and review of the standard being of most importance for the protection from corrosion in Germany.

Translations into English of the various sections of DIN
55 928 will soon be available.

6 Contract organisation: designer— constructor—supplier—inspector

J. V. BARTLETT, CBE, MA, FEng, FICE, FIEAust,
FASCE, and D. W. SMITH, BEng, MA, FICE,
FIStructE, FWeldI, MICorrT, Mott Hay & Anderson,
and G. F. MOBSBY, BEng, FIMechE, MIMarE, FIQA,
MIBF, J. Haggie Patterson & Associates

SYNOPSIS. Some shortcomings in present arrangements for
corrosion protection are listed, as are the difficulties
inherent in performance specifications. The paper seeks to
set out features which will increase the efficacy of a
protective system specified in a construction contract and
monitored and inspected throughout the performance of the
contract.

INTRODUCTION

1.   If a structure needs to be protected against corrosion,
the capital asset that is at stake is the structure itself.
It is by the value of this asset that the importance of
properly organised protection should be judged, and the
structural engineer must accordingly accept responsibility.
In an ideal world this would present no problem.

2.   But circumstances are not ideal. There are, of course,
splendid exceptions, but it is broadly true that the present
situation falls short in four distinct respects:-

i)   Many of us who are structural engineers either regard
     corrosion protection as a bore, or think we know it all,
     at least in our particular line, whereas in fact we are
     rather ignorant or out of date.

ii) Some corrosion technologists are over-idealistic, un-
     practised in the art of the economically possible.

iii) Some fabricating works are unable or unwilling to give
     corrosion protection its due importance. This is a
     matter of attitude, or organisation, and of physical
     provision. The economic pressure against improvement is
     formidable, except in those works which can expect a
     steadystream of "cream" jobs.

iv) Britain lacks a reliable system of selecting, training

and approving the right calibre of inspector.[1,2] Nor is
there any easy solution, since corrosion technology is
both multifarious and changeable. The inspector who is
good for this job today may not be so good for that job
tomorrow.

3.   As will be seen, none of these items concerns the technical question as to how protection can best be given.   It
has been well said that ".... adequate solutions exist for
most requirements, so that corrosion is a non-problem in a
technical sense".[3] Actually this is an understatement,
because for many requirements there is a whole range of
adequate solutions; and this is more often a curse than a
blessing, since time and effort are wasted in choosing between a redundant surplus of solutions of almost equal value.

4.   In the light of all this, it is clear that it is organisation and control that have to be improved, and we need to
ask both what can be done in the circumstances now reigning,
and also what changes in the circumstances can reasonably be
worked for.

TYPES OF CONTRACT

5.   The main divide between types of contract for corrosion
protection is that between contracts which specify a protective system and those which specify a performance.

Specification by performance is rare, and the question
must be asked, should it become more usual?

6.   The advantages of specification by performance are
obvious:-

a)   The client is relieved of a head-ache.

b)   There is an incentive on the applicator to do a good
job.

7.   The objections are a little less obvious, and require
to be spelled out more fully:-

a)   Applicators are reluctant to accept full responsibility.
The full cost of remedial measures in the event of a
failure may far exceed the cost of the original work,
and a contract which requires the applicator to bear it
all may prove impossible to arrange. If the applicator's
liability is limited, the client may still have his head-
ache.

b)   A long-term performance contract may prove difficult to
enforce. This head-ache can never be fully off-loaded,

first, because it may be difficult to prove, years after
the work was done, whether the contract has been met;
secondly, because the people on whom responsibility must
be fastened may, at the time of a failure, no longer
exist, or be difficult to identify, or be a consortium
that has broken up, or be no longer capable of fulfilling
their obligations. If, to minimise this risk, the duration
of the guarangee is made short, the definition of performance may be made even more difficult, and at best
the client's head-ache is merely delayed.

c)   The steelwork fabricator or erector may, for whatever
reason, act in a way which hinders the applicator from
doing a good job, and thus release the applicator from
his bond. This is the Achilles' heel of all initial
protection of structures, but it applies with especial
acuteness to the "performance-guarantee" contract. The
most common form of the difficulty is delay, in a harm-
ful environment, affecting steelwork to which only part
of the protective system has been applied. There is also
the problem of damage during transport, handling and
erection. The only ideal, from the client's point of
view, would be a steelwork firm which accepts the responsibility for successful protection; but this kind of
contract would be even more difficult to arrange.

d)   A complete structure, including its fitments, is likely
to require a variety of protective systems, involving
separate or subcontract applicators. An applicator may
quite reasonably be unwilling either to guarantee a
subcontractor's work, or to give an unqualified guarantee
even of his own work when other independent applicators
are involved.

8.   It is fair to conclude that the performance-guarantee
type of contract will always be the exception, rather than
the rule, in the general run of structural engineering.

     Specification by system:-

9.   The central question, then, is how to maximise the
effectiveness of protection by specifying the system.   The
main requirements, omitting ifs and buts, may be summarised
as follows:-

a)   The advice of a corrosion consultant should be sought
while the detailed design of steelwork is being done.
Ideally, the inspecting organisation is the appropriate
consultant; but unfortunately some of the best
inspecting organisations are unable to make themselves

really useful in this way; and so they will remain until given the opportunity to practise. (But good corrosion protection begins long before detailed design. It begins in the mind of the structural engineer when the design is being evolved.)

b)   The inspecting organisation should be invited to assist in selection of the various protective systems, and also in writing the specification, ideally by initiating a draft for discussion with the structural designer. If this degree of early involvement is to achieve its maximum benefit, it demands, but it also fosters, a degree of mutual comprehensive between the structural engineer and the corrosion consultant which is well beyond the current average. The question whether the inspecting organisation should be a department of the designer's own firm or authority is a matter of business convenience, as the pros and cons are evenly balanced from the point of view of achieving good protection against corrosion.

c)   The specification should be thorough, and should cover not only the technical requirements of all the protective systems to be used, but should also include a detailed inspection schedule, on the lines dealt with in section four of BS 5493.

d)   The final form of the specification should be prepared after the appointment of the structural firm. This means that the appointment of the steelwork firm should be made on the basis of a tender that is avowedly incomplete in certain important particulars. Though out of line with major areas of traditional British practice, this is the only way to ensure effective use of the resources of the particular steelwork firm. Of course, corrosion protection is only one of a number of activities to which this comment applies.

e)   The degree and methods of inspection should be appropriate to the needs of the particular project. In a structure such as a bridge, with a long design life and a high cost of access for maintenance, full inspection of initial protection is essential; and with any structure it is false economy to underestimate the degree of inspection needed. Whatever inspection is decided on, the appropriate amount of time, space and facilities must be allowed for. If for any reason the required standard of inspection is likely to be difficult to achieve, the question of using a more foolproof protective systems, such as hot-dip galvanizing, may have to be considered.

f) A contract that is based on controlling the system must control all aspects of the system, and that includes quality control of paint materials. Unfortunately, many paints are specified as proprietary produces, and specifications for control of quality are not available to the structural designer or the inspecting organisation. Too often the only enforceable clause in the contract dealing with the quality of paint is the name of the label on the tin. Paint manufacturers are of course entitled to their secrets, and it is understood that it is mainly for this reason that the new bridge code[4], which claims to be comprehensive, will not contain a part dealing with corrosion protection; a serious deficiency. But each paint supplier should be required, at the very least, to give details of two or three simple tests which the buyer can carry out as a check on consistency, such as density and viscosity, together with the tolerances within which the manufacturer intends to work to. This will at least help to eliminate mistakes.

A further major step was taken (for example) in the superstructure contract for the Forth Road brodge. The paint manufacturer was required to supply, in confidence to an independent laboratory, a note of his formulations. A fairly searching set of tests was then agreed between the manufacturer and the independent laboratory, and submitted to the engineer for approval. Throughout the work, routine samples were subjected to these tests by the laboratory, and the engineer was informed of any deviations from initial approved samples. In some contracts (not the Forth Road bridge), important discrepancies have occasionally been dealt with this way. The manufacturer's secrecy was fully and permanently respected, and further contracts followed for materials which could not be copied. Such a procedure is perfectly applicable in contracts of moderate size as well as large ones, and would be a useful addition to the repertoire of control. Whichever way is followed with regard to laboratory tests on material supplied, there should also be a very simple system for testing materials at site (eg density), which can be carried out quickly on samples taken both from the site store and from paint kettles.

10. For the purpose of this paper, the essential points to be emphasised are that these measures must be set out with complete definiteness in documents which are binding under the contract; that they involve expense not only to the inspecting organisation and the applicator, but also to the steelwork fabricator and erector; and that therefore the prices of these bodies cannot be complete until these documents are ready for signing.

ORGANISATION AT THE FACTORY

11. It is both usual and very desirable for as much as possible of the protective system to be applied at the factory before despatch to the site. Normally this means application in the fabricator's works, though for a very large structure a separate factory devoted entirely to corrosion protection can be used with great success.[5]

12. If protection at the fabricator's works is to be successful, the fabricator must provide adequate space for both the work and the inspection to be carried out, together with the necessary equipment and control.

13. A specimen clause is given in the Appendix stating the equipment for quality control which the contractor must provide.

14. The space provided should be under cover, and be well ventilated. It should be served by the largest overhead travelling cranes in the works, since the largest fabricated components are to be handled there. At least two segregated bays are needed, respectively for surface preparation and coating. They should be so located that fabricated parts can readily be brought in from the fabricating shop, and after coating, can readily be loaded onto vehicles for transport to site. To allow time for all the required operations, including drying of paint between coats, the protection bays ought to be larger than is at present usual by comparison with the fabrication bays. The programme should allow reasonable time for occasional remedial action to be taken before despatch of steelwork to site.

15. The work in the protection bays requires to be organised with the same imagination, skill and enthusiasm as work in the fabrication bays. Failure to organise the protection bays

efficiently is to create a traffic jam in the movement of fabricated steelwork, causing delay to the entire project. The organisation of the protection bays includes arranging time and handling for the work of protection, quality control and inspection.

16. If the corrosion protection is necessary at all, it is the structure that will eventually suffer if the protection is not properly done. The fabricator should take seriously the need to control the quality of the work. This involves provision of the right standard of supervision, and provision and maintenance of the necessary equipment and instrumentation. To neglect quality control on the grounds that protection is subsidiary to steelwork is equivalent in foolishness (for example) to a banker who seeing that locks are less valuable than cash, does not bother to provide a reliable lock for his strong-room door.

17. Steelwork fabricators vary enormously in the excellence and scale of provision they make for the application and control of protection. It is for engineers to ensure that contracts are awarded to fabricators who provide and operate the necessary facilities. Tender documents should indicate what is required, and require each tenderer to submit a full statement of what he can offer. The engineer should inspect the works of the fabricator whose tender he proposes to accept, and ensure that the contract, in its final form, obliges the fabricator to take steps which not only meet the specification, but which also are known to be within his capability as to time, type and quality of work.

18. It is also important for engineers to specify comparable factory protection regularly, so that fabricators have a sufficiently regular flow of work to justify maintenance of the required facilities and organisation.

SITE ORGANISATION

19. Work at a civil engineering site is not dependent on fixed capital assets or a permanent organisation in the same way as applies at the fabricating works. Protection work should be undertaken by a subcontractor for whom the structural erector, or the main contractor, assumes responsibility. The subcontract should state the required quality control and inspection procedures, and define the equipment, instruments and site laboratory space to be provided. The minimum facilities would be similar to those required for factory inspection (see Appendix).

CORROSION IN CIVIL ENGINEERING

20. The time allowed in the programme should be adequate for all the necessary operations, and should include an allowance for remedial measure in case of occasional failure to meet the standard.

ECONOMICS

21. Decisions about corrosion protection should be taken on the basis of hard economic facts. The important question that must always arise is "How far should a costly initial system of protection be applied in order to keep down subsequent maintenance?" Corrosion engineers may argue with faith and passion that it does pay to pay now, on initial protection, rather than later on maintenance; they must learn to support this by clear economic reasoning based (for example) on discounted cash flow and the concept of present value.

22. It is easy to justify a high standard of initial protection when interest rates are low, but high interest rates tend to reverse this conclusion, especially if inflation is left out of the equation.

23. Every one who took out a mortgage 15 years ago knows that inflation has made paying off easier. But dogma among economists has generally led them to ignore inflation in the calculation of discounted cash flow; right when dealing with the distribution of the finite national cake; surely wrong when advising a small organisation whose borrowing more or less for initial protection will be too slight to affect the market. Really good initial protection of a normal structure cannot at present be justified on the basis of this strange economic thinking; and a further misleading element is the separate budgeting for major capital works on the one hand and maintenance on the other.

24. Corrosion and structural engineers may in future have nothing to organise together unless they acquire the habit of applying ruthless logic to the economic questions that affect them, and the skill to debate these issues on equal terms with economists.

CONCLUSIONS

25. A performance-guarantee contract has attractions, and may have the advantage if it can be arranged with a substantial corrosion protection firm, which can be relied on to fulfil its obligations even after a lapse of years.

26. The more usual contract must continue to be one which

specifies the system of protection. The technical specification in such a contract must cover all the systems to be
used on the entire structure, and must specify them
thoroughly, in a manner which can be enforced both as to
materials and as to workmanship. The schedule of inspection
must also be fully set out as part of the contract. The
programme must be such that the work and the inspection
involved in it can be efficiently executed within the time
allowed.

27. The fabricating shops should be so organised as to
ensure that proper facilities are available for corrosion
protection, and proper quality control applied. If necessary,
the final details of the contract should be negotiated after
the steelwork firm has been selected and the available
facilities are known.

28. Organisational questions should be taken into account
when choosing a protective system. If a full and reliable
organisation cannot be arranged a relatively fool-proof
system should be adopted.

29. Engineers must know how they stand with regard to
expensive initial protection of structures when the cost of
borrowing is high.

REFERENCES

(1) BS 5493, Code of practice for Protective coating of
    iron and steel structures against corrosion,
    Section 40.

(2) Correspondence in Bull.Instn.Corr.Sci.Tech., 1978,
    (68) 19, (69) 16, (70) 16.

(3) Whiteley, P., Corrosion : a non-problem?
    Struct.Engr., 1972, 50 (Dec) 478

(4) BS 5400, Steel, concrete and composite bridges

(5) Anderson, J.K. et al, Forth Road Bridge, Proc.Instn.
    Civ.Engrs., 1965, 32 (Nov), paras 4.123-129

CORROSION IN CIVIL ENGINEERING

Suggested specification clause
outlining minimum quality control facilities

Protective coating (production quality control facilities)

The Contractor or his sub-contractor shall have available the
basic quality control equipment indicated below at every
location where protective treatment is carried out:

Environment
Air thermometer (max/min)
Contact thermometer
Hygrometer

Paint Testing
100ml metal pyknometer (mass/vol cup) for
determination of density to BS 3900 : A12
75mm dia flat bottomed dish for determination
of volatile matter to BS 3900 : B2
Sampling dip cylinder and/or tube
Flow cup of appropriate size for determination
of flow time to BS 3900 : A6
Stopwatch
Stand with levelling screws and spirit level
100ml graduated measuring cylinder
Sieve BS 410 : 120 mesh
Thermostatically controlled air drying oven
Thermometers - immersion
Desiccator
Analytical balance 0-500 gm
Balance 0-10 Kg

Preparation
Reflectometer, Elcometer Surclean Model 153
Swedish Standard SIS 05 5900 pictorial
standards
Potassium ferricyanide test papers

Application
Wet film thickness gauges
Dry film thickness gauges - inspection grade
(Elcometer Inspector Gauge Model 111)
Dry film thickness gauge - precision grade
(Elcometer Minitector Model 150 FN)
Holiday detector

The equipment should be suitably housed in a laboratory and
with personnel proficient in its use able to carry out
routine quality control checks during the course of production.

7 Cathodic protection in civil engineering

M. D. ALLEN, MIM, MICorrT, and D. A. LEWIS,
Bsc(Eng), MICheme, FICorrT, Spencer and Partners

SYNOPSIS. Cathodic Protection in Civil Engineering (a brief
explanation of the principles, application and costs of
cathodic protection illustrated by descriptions of three
recently engineered schemes, viz: a water supply trunk main,
a natural gas pipeline in the Oman, an airport hydrant
refuelling system).

INTRODUCTION

Cathodic protection is an electro chemical method of
preventing metallic corrosion in aqueous solutions. In civil
engineering, the metal is inevitably steel and the aqueous
solution usually soil, fresh water or sea water. The
technique is commonest, therefore, for piling, offshore and
harbour structures, buried pipelines and ships. In this
Paper, the theoretical background is outlined and the
principal features of practical schemes illustrated in
recent projects completed in the field of pipelines.

CORROSION PROCESS

Metals which are produced by winning from the naturally
combined state, have a tendency to revert to this state
under the influence of the forces of nature in the form of
water and oxygen. Viewed by the Electrochemist, the reaction
is represented as an <u>anodic</u> reaction thus :

$$M \text{ (insoluble, metallic)} \longrightarrow M^+ \text{ (soluble,} \quad) + e^-$$
$$\text{(form of metal} \quad) \qquad \text{(combined metal)}$$

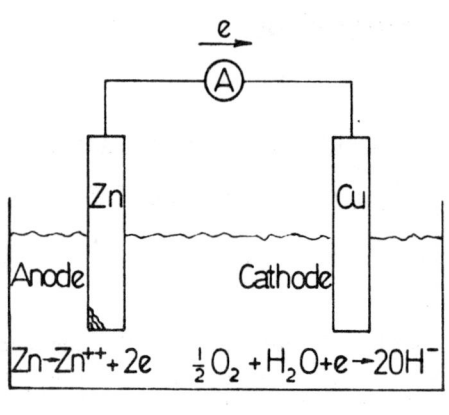

BIMETALLIC COUPLE

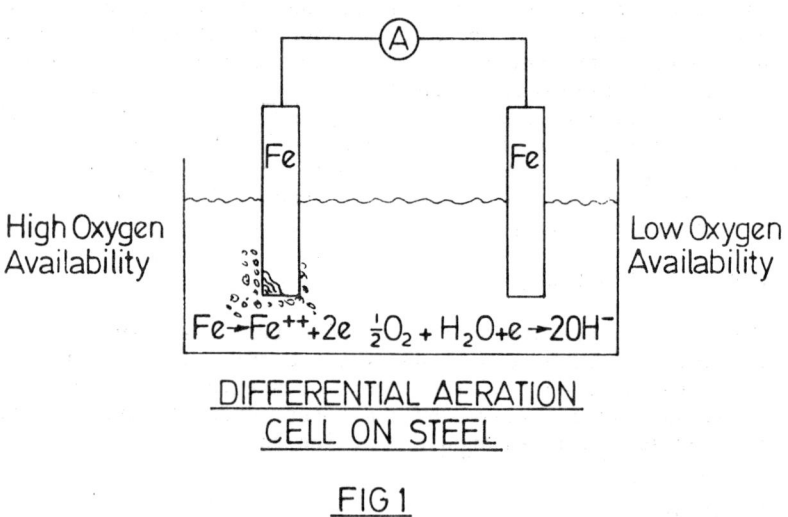

DIFFERENTIAL AERATION
CELL ON STEEL

FIG 1

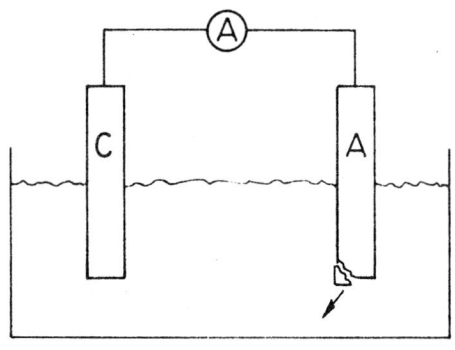

CORROSION CELL

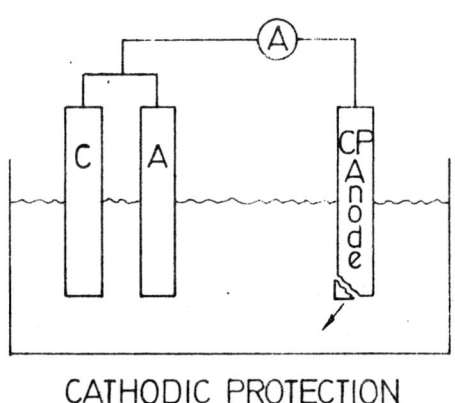

CATHODIC PROTECTION

FIG 2

CORROSION IN CIVIL ENGINEERING

In the transition from the combined to the metallic state, an electron is produced. Since the electron cannot exist on its own, some other simultaneous complementary reaction must occur. For example, the electron may combine with water and oxygen in a <u>cathodic</u> reaction thus :

$$\tfrac{1}{2} O_2 + H_2O + 2e^- \longrightarrow 2OH^-$$

(water and oxygen)   (hydroxyl ion)

These two reactions occur, for example, at the electrodes of the well known bi-metallic couple. However, a single metal such as iron can produce the same effect, since the tendency for iron to revert to the combined state will vary from location to location and is governed by variations either in the state of the metal itself or by differences in the nature of its immediate environment. Differentials in, say, heat treatment in the metal or temperature, or aeration of the environment can produce a series of anode/cathode reactions at different points.

The illustration of the corrosion process (Fig.1) shows a discrete separation of the anode and cathode: this need not be the case. Anodes and cathodes produced by differentials may also occur on the same steel article, separated either only by microns or by kilometres !

The principle of cathodic protection is to make all anodes on the metal to be protected behave as cathodes and this is illustrated in Fig.2.

CATHODIC PROTECTION

Referring to Fig.2, the auxiliary anode may be a metal less noble than iron so that the driving voltage of the cell becomes the difference in electrode potential between the two metals. Light metals such as magnesium, aluminium and zinc are commonly used in this application which is known as 'sacrificial anode protection'. Alternatively, the auxiliary anode may be an expendable or insoluble material which is energised by the application of a direct current, in which case the method is known as the 'power impressed system'.

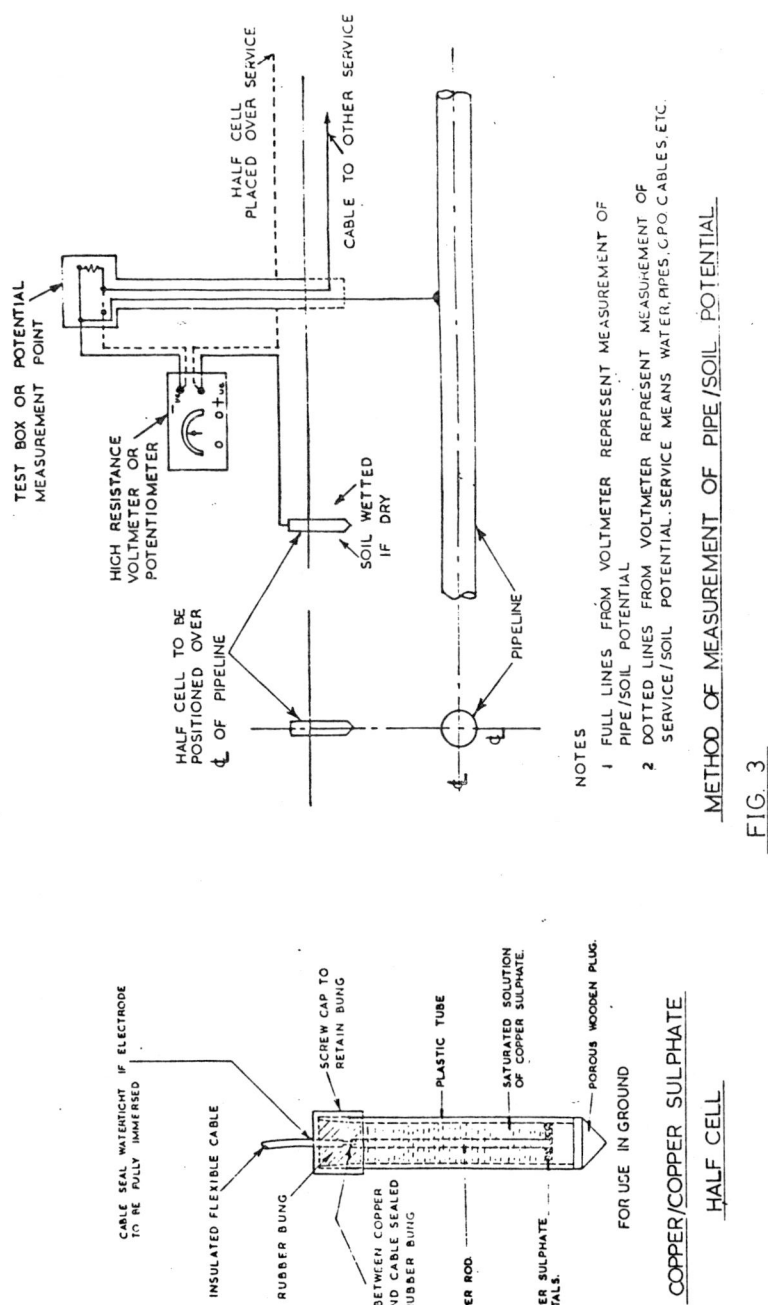

TEST BOX OR POTENTIAL MEASUREMENT POINT

HIGH RESISTANCE VOLTMETER OR POTENTIOMETER

HALF CELL TO BE POSITIONED OVER ℄ OF PIPELINE

HALF CELL PLACED OVER SERVICE

CABLE TO OTHER SERVICE

SOIL WETTED IF DRY

PIPELINE

NOTES

1 FULL LINES FROM VOLTMETER REPRESENT MEASUREMENT OF PIPE/SOIL POTENTIAL

2 DOTTED LINES FROM VOLTMETER REPRESENT MEASUREMENT OF SERVICE/SOIL POTENTIAL. SERVICE MEANS WATER, PIPES, GPO CABLES, ETC.

METHOD OF MEASUREMENT OF PIPE/SOIL POTENTIAL

FIG. 3

CABLE SEAL WATERTIGHT IF ELECTRODE TO BE FULLY IMMERSED

INSULATED FLEXIBLE CABLE

RUBBER BUNG

JOINT BETWEEN COPPER ROD AND CABLE SEALED INTO RUBBER BUNG

COPPER ROD

COPPER SULPHATE CRYSTALS.

SCREW CAP TO RETAIN BUNG

PLASTIC TUBE

SATURATED SOLUTION OF COPPER SULPHATE.

POROUS WOODEN PLUG.

FOR USE IN GROUND

COPPER/COPPER SULPHATE

HALF CELL

CORROSION IN CIVIL ENGINEERING

It can be shown in theory from a consideration of the solubility of iron, and has been demonstrated in practice, that the transition from metallic to soluble iron is prevented when the electrode potential of the iron object is depressed from its naturally occurring value to − 850 mV referred to a copper sulphate standard reference electrode. This electrode potential is the means by which it is determined whether the metal is protected. In practice, the potential is known as the "pipe to soil" potential and is measured as illustrated in Fig.3.

The amount of current necessary to produce this change in potential reduces significantly when an insulating layer is applied to the steel surface. Thus, while an uncoated pipeline may require a current in the order of 1 milliamp per square metre of metal to produce this change, coated steel may require only 0.1 milliamp per square metre. Since this represents a very significant reduction in total power requirement, it has become common practice to utilise cathodic protection as an adjunct to a sound insulating coating.

The current being provided to the structure must be confined to that structure and not dissipated to associated bare metal areas such as electrical earthing systems. In pipelines, insulating flanges are used for this purpose.

A power impressed system consists of a power source to provide dc current and a means of leading this current through the earth to the structure to be protected. A typical cathodic protection station is illustrated in Fig.4. It can be seen that since the protection operates by current flow in the ground, there is always the possibility that some of this current may pass to other structures in the vicinity. This is a well-known phenomenon known as interaction which can be minimized by careful design and overcome by bonding (Fig.5).

Cathodic protection should not be regarded as automatically necessary. Corrosion prevention facilities require both capital and maintenance expenditure. It is possible to make an assessment of the costs of allowing some corrosion to occur and the cost of various preventative measures.

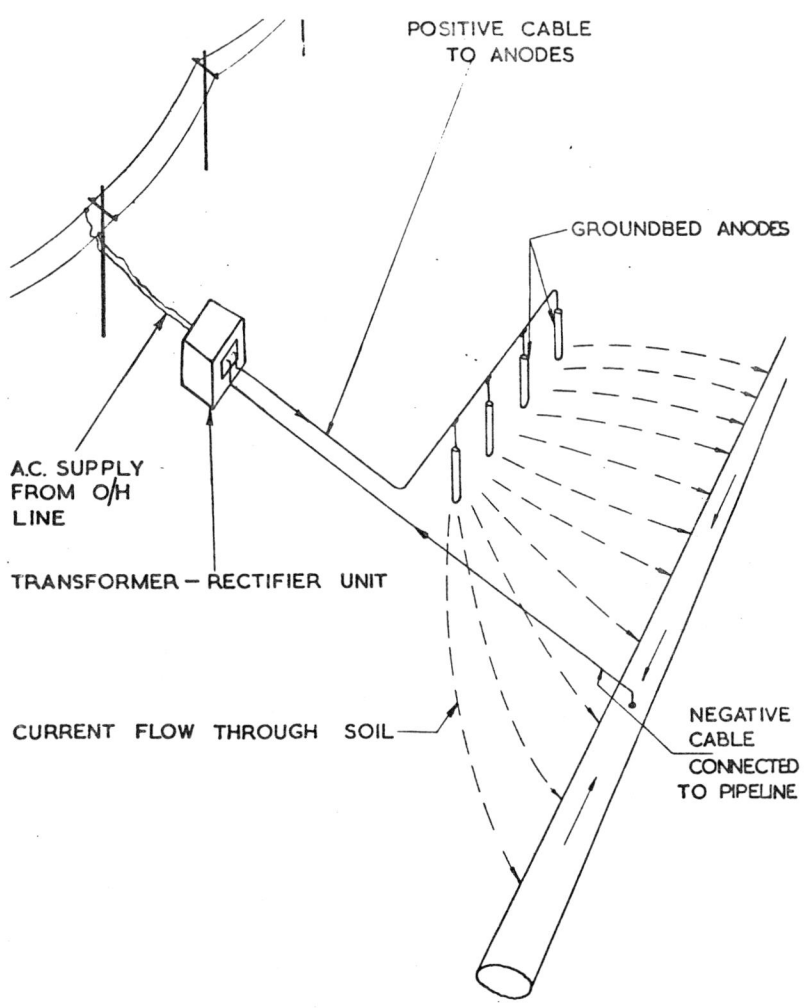

POSITIVE CABLE
TO ANODES

GROUNDBED ANODES

A.C. SUPPLY
FROM O/H
LINE

TRANSFORMER — RECTIFIER UNIT

CURRENT FLOW THROUGH SOIL

NEGATIVE
CABLE
CONNECTED
TO PIPELINE

TYPICAL POWER IMPRESSED CATHODIC PROTECTION
FIG 4

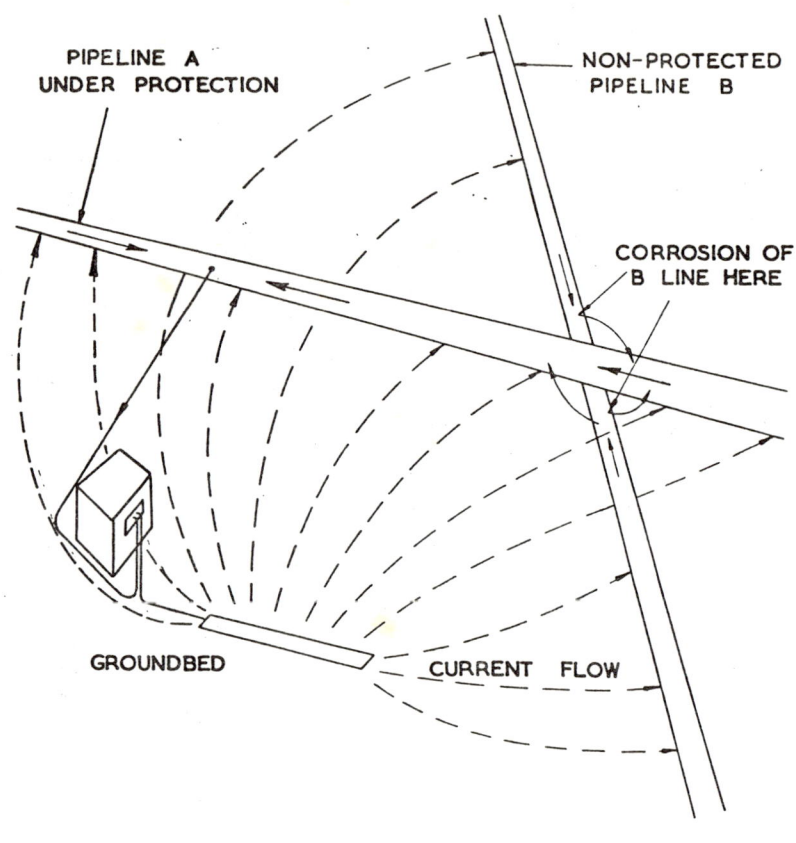

PIPELINE A
UNDER PROTECTION

NON-PROTECTED
PIPELINE B

CORROSION OF
B LINE HERE

GROUNDBED

CURRENT FLOW

INTERACTION AT PIPE CROSSING
FIG 5

In the case of harbour structures and water pipelines, for example, full life expectancy may be more readily achievable by increasing the metal thickness or undertaking repair work. It is only in the case of high pressure gas pipelines or pipelines carrying flammable or noxious liquids, that complete corrosion prevention is required on the grounds of safety or environmental protection.

In the examples of cathodic protection applications which follow, the various principles given above are illustrated for power impressed schemes on pipelines.

A NATURAL GAS TRANSMISSION LINE

In 1977 a 20 inch diameter natural gas line, 325 kilometres in length was built in the Sultanate of Oman. Prior to construction, a corrosion survey had shown that the gas line route traversed some aggressive soil conditions. There were also areas of consistently high resistivity non-aggressive soils, where the corrosion hazard to mild steel would be expected to be only very slight. However, for long welded steel pipelines running through soil conditions which vary considerably, a complete coating and cathodic protection scheme was considered essential since no perforation by corrosion was acceptable.

The routing of the gas line was complicated by its close proximity, over some 225 kilometres of its length, to an existing oil line, with the two lines actually crossing at three locations.

The oil line had been commissioned some ten years previously and had also been provided with a cathodic protection system, consisting of eight stations. Due to a high ground-bed resistance, this system was unable to provide its full current output capacity and levels of protection on the oil line were borderline. The operation of these other cathodic protection stations would be expected to cause interaction currents to be associated with the gas line, creating a further significant corrosion hazard. A cathodic protection system incorporating direct and/or resistive cable bonds between the oil line and the gas line would overcome this hazard.

CORROSION IN CIVIL ENGINEERING

It was a prerequisite of the gas line protection scheme
that it should also reinforce the protection of the oil
line where possible. In fact, an integrated scheme, where
both the new stations on the gasline and the existing
stations on the oil line, were designed to supply current
to both lines, was found to be more economic than
independent schemes for the two lines.

The gasline coating system consists of a factory applied
sintered epoxy coating 300 microns thick with field joints
coated using an 875 micron thick PVC laminated tape and
primer system. A sintered epoxy coating is, outside of the
USA, a relatively unknown quantity as far as its behaviour
characteristics under cathodic protection is concerned.
However, it was considered from existing knowledge that
current demand would be as low or lower than the conventional reinforced coal tar enamel coating. It was, however,
considered advisable to design to a maximum impressed
negative potential of - 1.5 volts.

In the design it was possible to consider five means of
supplying the necessary dc power. These were, thermo
electric generators, solar panels with lead acid batteries
and transformer-rectifiers coupled to either diesel or
natural gas generators or in some cases, an existing ac
mains supply. The provision of dc cathodic protection
supplies from transformer-rectifiers, associated with an
existing secure ac mains supply is comparatively inexpensive,
reliable and very low on maintenance costs. Equally reliable
and requiring little maintenance are solar panel installat -
ions. However, for station outputs of greater than 3 or 4
amps, capital costs are presently very high. With a constant
supply of natural gas from the pipeline, thermo electric
generators and gas engine generators would have a high
degree of reliability. For a power requirement of up to
250 watts, thermo electric generators connected to a secure
supply of natural gas would have a definite cost advantage
over both forms of transformer-rectifier/generator units.
For diesel generators, the cost of transporting fuel and
maintaining the engines, had also to be considered.
Maintenance costs of the engines is also a consideration,
although this is a less significant factor for natural gas
generators. It was decided by the pipeline operator that
the provision of the small off-takes for the gas supply to

cathodic protection stations were unacceptable and bearing
in mind the desirability of standardizing equipment throughout the scheme, the final selection was for transformer
rectifiers with dual automatic diesel generator sets which
could run unattended for up to 45 days.

For safety reasons, dc output of the transformer-rectifiers
was limited to a maximum of 50 volts. With groundbed design
based on a 1 ohm circuit resistance, some 50 amps can be
realised from each station.

Groundbed design varied with the prevailing ground conditions.
With ground resistivities below 5000 ohm-cm at 1-3m depths
it is usually economical to construct a horizontal groundbed.
Where these resistivities are only found at depths of 3-6m,
vertical anode groundbeds can be utilized. Where it is not
possible to utilise either of these designs, deep (borehole)
groundbeds have to be employed.

Due to varying soil resistivities along the route, where
ground conditions were found to change from low resistivity
gypsum soils to very high resistivity mountainous rock, all
three types of groundbed were employed at various locations
along the route.

The scheme as installed comprised six new stations, the
improvement of two existing stations to realise their full
output capacity, and minor modifications to a further two
existing stations to enable current to be supplied to both
lines.

The materials for the scheme cost come £70,000 delivered to
the Oman and installation costs were approximately a
further £150,000.

Operation and maintenance of the scheme, by the Owner will
include the measurement of pipe to soil potentials and bond
currents at least every four months by corrosion technicians,
supervision and analysis of these results by a Corrosion
Engineer and maintenance and refuelling of the diesel
generators. It was estimated that, were all these services
provided by independent Contractors, costs for operation
and maintenance of the system could amount to some £18,000
per year.

CORROSION IN CIVIL ENGINEERING

AN URBAN TRUNK WATER MAIN

In planning a 36 inch diameter trunk water main in London, the choice of materials for construction included steel provided with cathodic protection.

Through such heavily built-up areas, numerous other services were encountered and the design of the coating and cathodic protection was therefore carried out to minimize current demand and therefore interaction effects. The main was constructed with spigot and socket welded joints. An internal lining of bitumen was provided and the external coating was bitumen enamel, reinforced with fibre glass.

As is usual with large water mains, numerous special fitt - ings and line valves were installed, each valve being provided with bypass and washout facilities. Each main line valve and each off-take or bypass valve was provided with an electrically insulated flange. In the case of the off-take and bypass valves, the installation of these electrical discontinuities ensured that the uncoated auxiliary pipework which constituted unacceptable current earthing, was isolated from the pipeline and the protection of the main line thus preserved at minimum current. The provision of main line electrically insulated flanges allowed the cathodic protect - ion current to be controlled by cable bonds to each section. Special attention was also paid to the coating of all valves, flanges, offtakes and pipework passing through valve chambers to minimise overall current demand.

Cathodic protection of the main has been achieved with a single impressed current system, supplemented with magnesium anodes at specific locations of anticipated higher-than-average-current demand. Current is provided from a single horizontal groundbed of 30 metres length consisting of silicon iron anodes surrounded by coke-breeze, installed approximately midway along the pipeline route. The ground-bed is energised by an oil immersed transformer-rectifier connected to the local Electricity Board supply. Maximum capacity is 24 amps 24 volts. As is normal in the UK, this transformer-rectifier was pole mounted to avoid the con-struction of a security compound. Over the total length of the main, over eighty concrete posts were required to be installed to house the various test and bond cables

required for interaction testing, remedial bonding and
routine monitoring of effectiveness.

The system was commissioned with a resistive bond across
only one of the main line insulating flanges, restricting
current to the Northerly five kilometres of trunk main.

The current demand of the trunk main was found to average
0.1 milliamps per square metre of pipe surface area at the
time of commissioning in 1974. Over the last four years the
current demand has increased by approximately 20%.

At the time of commissioning, interaction testing was
carried out with all Authorities concerned who had metallic
services crossing or close to the trunk main. Only one
location was found where interaction effects exceeded the
limits recommended in the BS Code of Practice 1021: 1973.
Thus the designed provisions to minimise the output of the
cathodic protection unit significantly reduced the need for
costly remedial works to overcome interaction.

Maintenance monitoring of the cathodic protection system is
carried out over two days every six months, including the
measurement of all pipe to soil potentials and bond currents
and checking the operation of the transformer-rectifier.
Repeat interaction tests are offered every year to all
Authorities concerned.

AN AIRPORT REFUELLING SYSTEM

Some twenty years ago, two major Petroleum distribution
companies each constructed five pipelines of between 4 inch
and 8 inch diameter to carry petroleum products from their
Perimeter Tankage at London (Heathrow) Airport to Depots
adjacent to the Central Area Airport Terminal Buildings
from where fuel was loaded onto road tankers for transfer to
aircraft. The lines were each about 1200 metres, of all
welded steel construction and were coated with asphalt
enamel. A corrosion survey, consisting of the measurement of
ground resistivities at depths of 1, 2 and 3 metres and
analysis of soil samples had indicated that ground conditions varied along the route from occasionally "aggressive"
to being, more generally, "unlikely to be aggressive". At
the time, it was decided not to proceed with cathodic
protection. During 1967, evidence of external corrosion on

some of the pipelines was found. It was then decided that
the protection of the lines should be reinforced by a
cathodic protection system to avoid further corrosion as the
expansion of the Airport necessitated the completion of
further taxiways over the pipelines making any repair of
these lines both expensive and disruptive to Airport
operations.

During the feasibility study, it was found that suitable
flanges could be located for all the pipelines to be
insulated at appropriate locations and a current drainage
survey demonstrated that some 12-15 amps would be required
for protection of all ten lines. It was assumed that as the
pipelines were already ten years old, little additional
current allowance needed to be made since no further
significant natural coating deterioration was likely to
occur.

A major problem was the numerous metallic 'foreign' services
which traversed the pipeline routes and it was expected
that a further 10-15 amps might possibly be required to
overcome interaction effects. Because of the probable
interaction problems, a totally magnesium anode cathodic
protection scheme was considered. This was discounted
because of the unsuitable ground conditions over the
majority of the route for these anodes and because such a
system would require multiple excavations along a pipe
corridor which crossed operating runways and taxiways.
However, before a power impressed scheme could be proposed,
it was necessary to secure the agreement of all the fuelling
companies to a single shared scheme.

It was decided to install an oil cooled transformer-
rectifier, with a borehole type groundbed. The transformer-
rectifier would be capable of providing enough current for
both present and projected future pipelines. A borehole
groundbed requiring a minimum of ground space, was located
in an existing perimeter fuel depot and was expected to
further reduce interaction effects.

The scheme was managed on behalf of some eleven different
Petroleum Companies, independently and in various consortia
by one of the companies and the design allowed for the
measurement of power consumption in each operator's circuit.

In 1968-69, the protection was extended to cover the then
new hydrant aircraft re-fuelling lines for the long-haul
piers and Cargo Area and further pipelines from the
perimeter to the Central Area.

Because of the protracted negotiations concerning cost
sharing, operational responsibility and interaction
provisions, installation of the scheme covered a period
of some two years, eventually being commissioned in 1971.

The scheme as commissioned, consisted of a transformer-
rectifier with a maximum capacity of 50 amps at 48 volts
with a twin borehole groundbed system of silicon iron
anodes in graphite at some 50 metres depth. Current was
distributed to the various pipelines through a metered
negative distribution panel, where by the insertion of
fixed resistors into each circuit, the amount of current
for each pipeline could be controlled and measured.

The measurement of pipe to soil potentials on the Airport
is complicated by the fact that 90% of the pipelines in the
system are under apron, taxiway or runway concrete. To
overcome this difficulty, the majority of the readings are
taken at the hydrants, vents or drain points, where
electrical connection to the fuel mains is possible.
Installed in the concrete aprons, close to selected hydrants,
etc, are vertical 100mm diameter pitch fibre pipes which
permit the reference electrode to be located in the soil at
approximately pipe invert level, the top of the access point
being covered by a standard valve cover.

Since commissioning of the scheme, the hydrant fuelling
system at the Airport has expanded continuously. Apart from
the original ten lines, there now exist some 30km of pipeline, ranging from 6 - 24 inch, all provided with a
reinforced coal tar coating. Current demand for the original
asphalt coated ten lines remains some 10 amps. For the
30km of newer coal tar coated pipe, some 6 amps provide full
protection.

The scheme cost in 1971 was some £27,000 for materials,
installation, supervision and commissioning. Maintenance
costs are approximately £1500 annually, covering two visits
per year by a Cathodic Protection Engineer to measure pipe

to soil potentials, check insulated flanges, measure the various pipeline currents and carry out repeat interaction tests. One of the Petroleum Companies accepts operating responsibility but the running and maintenance costs are shared by all the various Companies, the percentage share being calculated from the actual proportion of the total cathodic protection current used by each Company for the particular pipelines they own and operate.

CONCLUSION

These examples were intended to illustrate various aspects of cathodic protection scheme design and in particular to emphasise the relationship between the coating of a structure and cathodic protection current demand. The overall cost of a cathodic protection scheme is directly related to the amperage required; however, dealing with the effect of the scheme on 'foreign' structures can be a significant proportion of the cost of the protection of the primary structure. The quality of coating should be such that the system may operate at an output unlikely to produce interaction and groundbeds should be located and sized to minimize local interaction effects. Installed systems must be provided with regular maintenance and monitoring to ensure continued protection.

8 Weathering steels

M. B. KILCULLEN, PhD, BSc, MICorrT, British Steel
Corporation, and M. McKENZIE, MSc, BSc,
Transport and Road Research Laboratory

SYNOPSIS

The corrosion performance of Cor-Ten A and Cor-Ten B
steels was measured at a range of sites in the UK and the
results of this work are presented. The case for using
Cor-Ten steel in structures is examined and it is
concluded that under certain circumstances the use of
Cor-Ten steel can be an economically attractive
proposition.

INTRODUCTION

1. In the late 1920's and early 1930's the United
States Steel Corporation, working on the effect of
alloying elements on the properties of steel, produced a
new steel composition which they called Cor-Ten. It was
developed and used mainly for its increased strength as
compared with mild steel but its corrosion performance
when exposed bare to the atmosphere was also found to be
much superior to mild steel. Over the years there were
many reports of its beneficial performance culminating
in 1961 in a major report by Larrabee and Coburn[1] which
showed that a Cor-Ten steel suffered a loss of only 40
um during 15.5 years exposure at an industrial site, as
compared with 0.72 mm loss by mild steel.

2. These and other results convinced American architects and engineers that this steel could be used
without protective coatings and, following the
development of a structural grade of the material, the
world's first structure, using uncoated Cor-Ten steel,
was completed in 1963. With this building, for Deere
and Company at Moline Illinois, the architect Eero
Saarinen gave birth to the concept of Weathering
Steels. In the years that followed most major steel
companies in the world produced their own weathering
steel or produced Cor-Ten steel under licence from U.S.
Steel Corporation. In this country a number of steel
companies, took out licences to make Cor-Ten Steel.

3. Some work had been carried out in the U.K. on this
type of steel^{2,3} and it was obvious from the results of
this that under suitable conditions of exposure the
corrosion rates of Cor-Ten type steels were much lower
than the corresponding rates for mild steel. These
results, however, also showed that the rates observed in
the U.K. were considerably higher than those being
reported from America, although the corrosion rate was
found to reduce with time. This shows that the same
basic mechanism is at work here and in the U.S.A. i..e.
the production of a rust film which develops protective
properties.

4. The protective properties of the rust film take
quite a long time to develop, in some cases many years,
and the extent to which they develop depends very much
on the atmospheric conditions of exposure. Consequently
it is very difficult to predict the long term
performance of these steels even on the basis of tests
carried out over a number of years.

5. The rusts formed on these steels in either soil or
water are not protective in any way and corrosion rates
similar to those experienced with mild steel are
obtained.

6. With the increasing interest in this country in the
use of Cor-Ten a considerable amount of work has been
undertaken by the British Steel Corporation and by the
Transport and Road Research Laboratory to investigate
the performance of Cor-Ten in U.K. climates. Both the
B.S.C. work, which is of a general nature and the TRRL

work which is aimed more specifically at highway bridges
is reported in this paper and the results and their
implications for the use of Cor-Ten for structures in
the U.K. are discussed.

Results of work carried out by British Steel
Corporation.

7. All the early work on Cor-Ten steel both in this
country and abroad had been done with the original
architectural Grade A material. A structural material,
Grade B, was developed in the late 1960's and it was
decided that work should be done on both Grades.
Samples of these materials and of mild steel were
exposed at the 11 sites listed in Table 1, and the
corrosion losses were measured after 1, 2, and 5 years
exposure at each site. It is not proposed to give all
the results in detail in this paper but rather to give
the results from typical sites which show the general
nature of the results. Consequently Figs. 1 and 2, set
out the results for Battersea, an industrial site,
Teesside a severe industrial site, Rye, a marine site
and Brixham, a rural site. To obtain a measure of the
decrease in corrosion rate with time, the rate of
corrosion which occurred during the first and last years
of the test at each site were calculated and these
results are given for Cor-Ten B in Table 1.

8. The results show that the corrosion rate of mild
steel is significantly higher than that of Cor-Ten B
which again was found to corrode more than Cor-Ten A
steel. They also show that the range of corrosion rates
of Cor-Ten steels was considerably narrower than the
equivalent range for mild steel.It was found that, from
all the sites involved in the test, the range of
corrosion rates during the last 3 years of the test was
9 um/year to 63 um/year for mild steel, 5 um/year to 33
um/year for Cor-Ten A steel and 7 um/year to 36 um/year
for Cor-Ten B steel.

9. In addition to these tests the corrosion rates of
Cor-Ten B steel were measured after periods of exposure
up to three years underneath the Cor-Ten steel bridge at
Iden in Sussex and, in conjunction with TRRL, under two
Motorway bridges, one at Cutbush Lane (M4) and one at
Sandiacre (M1). The results of these tests are given in

CORROSION IN CIVIL ENGINEERING

TABLE 1

CORROSION PERFORMANCE OF COR-TEN B STEEL -
FREE EXPOSURE - 5 YEARS

Site	Corrosion Loss in 1st Year of Test um	Corrosion Loss in Last Year of Test um
Battersea	56.3	27.6
Stratford	69.4	28.8
Teesside	67.7	26.3
Motherwell	55.0	22.3
Sheffield	70.0	35.3
Birmingham	75.0	35.8
Shoreham	52.0	9.1
Rye	52.5	13.8
Leatherhead	45.0	14.5
Brixham	31.4	7.1
Avon Dam	40.9	14.5

TABLE 2

CORROSION PERFORMANCE OF COR-TEN B STEEL -
SHELTERED EXPOSURE (UNDERBRIDGE) - 3 YEARS

Site	Subject to de-icing salt	Loss in 1st Year of Test um	Loss in Last Year of Test um
Iden Bridge	NO	22.1	17.5
M1 Motorway Bridge	YES	37.8	25.2
M4 Motorway Bridge		43.1	25.3

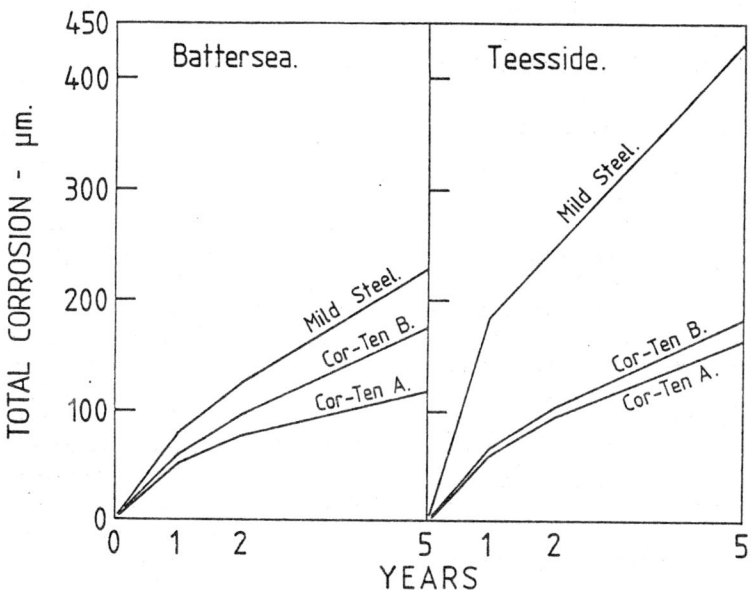

Fig.1   CORROSION - TIME CURVES FOR MILD STEEL,
COR-TEN 'A' STEEL, AND COR-TEN 'B' STEEL
AT BATTERSEA AND TEESSIDE.

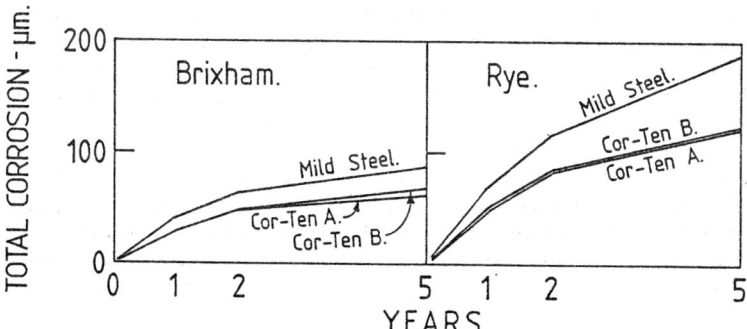

Fig.2   CORROSION - TIME CURVES FOR MILD STEEL,
COR-TEN 'A' STEEL, AND COR-TEN 'B' STEEL
AT BRIXHAM AND RYE.

CORROSION IN CIVIL ENGINEERING

Table 2, and they, once again, show the reduction in corrosion rate with time of the Cor-Ten steel although this is less marked than in open exposure.

Results of TRRL work

10. TRRL interest in Cor-Ten arose through its use in bridge construction in North America. Existing long term tests with weathering steels usually referred to open exposure conditions whereas much of the steelwork in a composite bridge is sheltered. Therefore in addition to open exposure testing at a range of sites, corrosion rates of Cor-Ten B were also measured under bridges near to some of the test sites. These bridge test sites were: Tinsley (Yorks) an industrial area next to Tinsley viaduct (M1), Loudwater (Bucks) - a rural area with nearby light industry, adjacent to Loudwater viaduct (M40), Eastney (Hants) a marine area in Langstone harbour. The underbridge corrosion rates for comparison with the Eastney open exposure specimens were measured on the Norfolk bridge at Shoreham by Sea, Sussex.

11. Fig. 3 shows the range of TRRL results obtained in open exposure compared with a typical result from U.S.A. These results fit quite closely with the B.S.C. results. Fig. 4, gives the corrosion rates of Cor-Ten B in open and sheltered exposure. These show that except for the Eastney/Shoreham comparison the sheltered corrosion rates are considerably lower than those measured with freely exposed specimens. However, at Loudwater, the semi-rural site, there was much less tendency for the rate of corrosion to reduce with time under sheltered conditions.

The open exposure rate dropped from 50 um in the first year to an average of 10 um in each of the last two, whereas the sheltered rate dropped from 20 um in the first year to an average of 15 um in each of the last two.

12. It is interesting to note here that the total corrosion losses at all underbridge sites with the exception of the marine site are in the range 75 to 100 um over a five year period and at those sites where the tests have not yet run for five years, extrapolation of the three year results would suggest that they will fall in a similar range.

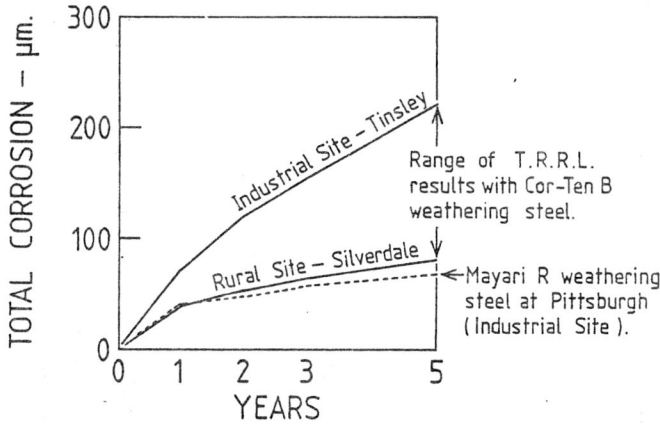

Fig.3   COMPARISON OF WEATHERING STEEL
        CORROSION RATES OBTAINED IN
        U.K. AND U.S.A.

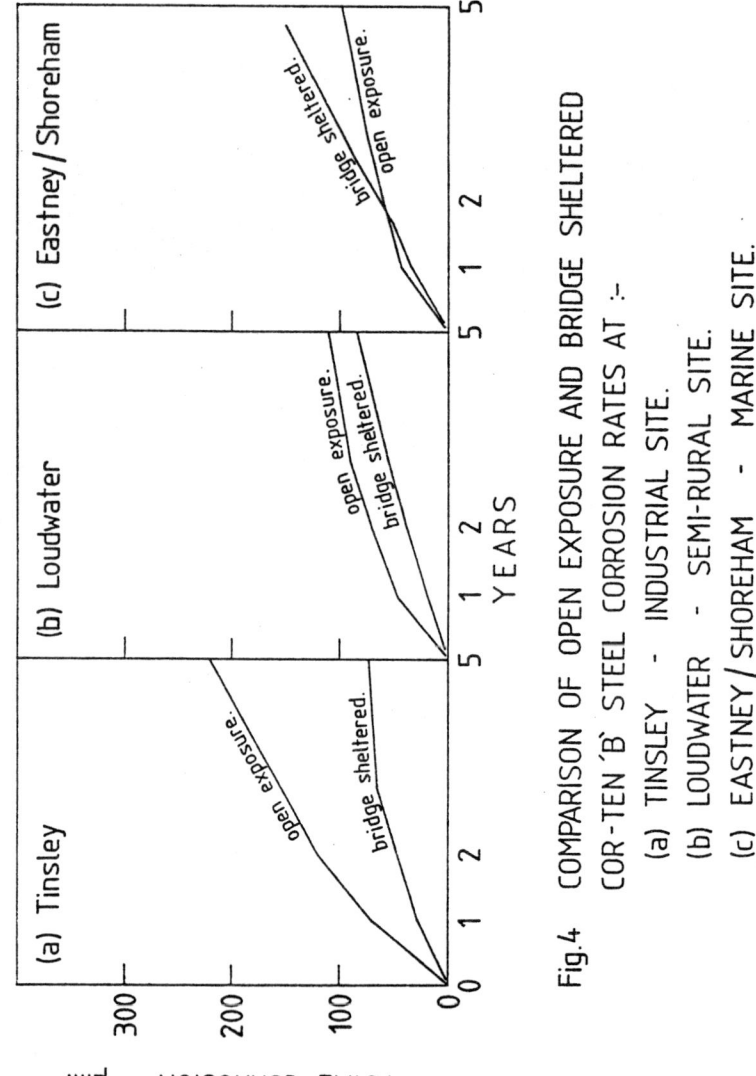

Fig.4   COMPARISON OF OPEN EXPOSURE AND BRIDGE SHELTERED
COR-TEN 'B' STEEL CORROSION RATES AT :-
(a) TINSLEY - INDUSTRIAL SITE.
(b) LOUDWATER - SEMI-RURAL SITE.
(c) EASTNEY/SHOREHAM - MARINE SITE.

13. In both the BSC and TRRL tests the corrosion rates were determined by weight loss measurement after the removal of all corrosion products in inhibited acid. As regards the appearance of Cor-Ten steels it is clear from the results of both tests that the colour and texture of the rust formed depends entirely on the conditions of exposure. Free exposure in industrial atmospheres produces a dark brown adherent rust whilst a rural environment produces a loose light brown/orange coloured rust. Sheltering tends to lead to light coloured loose rust films.

DISCUSSION

14. The work carried out by both establishments demonstrates two points very clearly. Firstly the corrosion rates of Cor-Ten steel are always significantly lower than that of mild steel and secondly the corrosion rates of Cor-Ten steels in the U.K. are higher than the results reported from America. However, having said that, it is the actual rate of corrosion of the steel that matters to the engineer and not its relationship to that of mild steel. The results show that Cor-Ten steels corroded at finite rates at all the test sites. The severe industrial sites were more aggressive towards Cor-Ten than any other site as regards open exposure but under shelter the corrosion rate, even at these aggressive sites, fell to levels not very much different from those measured at less aggressive sites e.g. Tinsley open exposure gave a corrosion loss of about 220 um as compared with about 75 um total loss under the bridge at the same site and 85 um total loss under shelter at Loudwater.

15. Contrary to what might be expected based on the American experience the performance of Cor-Ten freely exposed at or near the coast has been better than in industrial sites. There are however indications that where there is chloride contamination of sheltered sections of steel it can lead to some localised attack such as a mild form of pitting. This was observed both in the marine environments and on specimens on the undersides of the Motorway bridges where deicing salts are used in the winter months. It is still too early to

say how much of a problem this is and more work needs to be done in the area.

16. The case for using Cor-Ten must and should be based on economics and not on whether it performs well or badly in relation to a material whose use bare would never be considered. If, when used in a given application, the extra cost of Cor-Ten even allowing for corrosion, is less than what it would cost to protect a normal steel then its use makes economic sense.

17. Looking at all the data collected on the performance of Cor-Ten steel it is seen that only in the severest environments does its corrosion rate in the 5th year of exposure excede 25 um/year, and at a wide range of other sites it is considerably lower. Experience shows that the rate of corrosion will continue to drop beyond these levels before reaching a steady state so it is difficult to extrapolate with accuracy over long periods. It should be borne in mind here that the environment is changing all the time and future developments may alter the nature of a particular environment quite considerably. However until the results of work presently in hand to produce longer term data for Cor-Ten become available, designers will have to work on the basis of the most up to date data available. Consequently, it is recommended that, on any particular project involving Cor-Ten, the engineer contact the British Steel Corporation to obtain from them advice and information on the most relevant data available.

18. For many projects the use of Cor-Ten steel with a corrosion allowance, to postpone painting for 30 or perhaps more years would represent a considerable saving bearing in mind the cost of the initial painting and probably 3 to 4 maintenance repaints in that 30 year period. It is probable that with suitable monitoring of the structure the postponement of painting can be for periods considerably greater than 30 years or indeed may be for the whole life of the structure which of course, would mean greater savings still. There are many situations where the use of Cor-Ten steel with up to 1 mm extra thickness per surface allowed for corrosion will give a less expensive outcome than painting and maintaining ordinary steel even allowing for the possibility

that it may be necessary to paint the Cor-Ten at sometime in the future. In this way using Cor-Ten steel in this country can save money.

1. Larrabee, C.P., Coburn, S.K., Proc. 1st. Int. Congr. Metallic Corros. 1961 276, (London Butterworth).

2. Hudson, J.C., Stanners, J.F., J. Iron Steel Inst., 1955, 180, (3), 271.

3. Chandler, K.A., Kilcullen, M.B., Brit. Corros. J. 1968, 3, 80.

9 Protection of steel by metal coatings

F. C. PORTER, MA(Cantab), FIM, FICorrT, Senior
Officer, Zinc Development Association

SYNOPSIS. Steel - cheap, strong and versatile - is widely
used in civil engineering; metal coatings of zinc or
aluminium economically prevent the steel from rusting.
About 2,000,000 tonnes of zinc, 20,000 tonnes of
aluminium and 4,000 tonnes of cadmium are used each
year for coating steel throughout the world. Zinc is
applied mostly by hot dip galvanizing. Electroplating or
mechanical plating (of zinc or cadmium) and sherardizing
(with zinc) are used on smaller parts. Both aluminium
and zinc are applied by metal spraying. Details are given
of the properties of the metal coatings and examples are
quoted to show performance in practice and the economic
advantages.

INTRODUCTION

1.   Each year corrosion destroys the equivalent of one-
fifth of the annual world production of ferrous metals.
Most methods of protection involve a layer of corrosion
resistant material between the steel and the environment.
The main action of paints - and some metallic coatings
such as tin, chromium and nickel - is to provide a barrier
between the environment and the steel. However, if there
is damage to such a protective film, corrosion occurs at
the break and spreads outwards lifting and destroying the
film as it goes. Similar effects will also occur round
inherent pores in paint films. Zinc, aluminium and
cadmium not only provide an impermeable barrier between
steel and the atmosphere or solution but also protect steel

at small gaps in the coating.

2.   As the atmospheric corrosion resistance of a zinc
coating is proportional to its thickness and it corrodes at
a predictably slow rate (one-tenth to one-fiftieth that of
bare steel or even less in some marine environments)
zinc coatings can be specified to last many decades
without maintenance. This enables maintenance costs to be
greatly reduced and steel so protected is very economical,
as the first cost of coating steel with zinc is often no more
than the cost of a good paint system. Specifiers are
increasingly turning to zinc coatings also because they are
aware that the cost of keeping steel protected against
rusting by periodic painting is getting relatively higher all
the time because of the high cost of labour for maintenance
work and the high indirect costs associated with inter-
ruption of service for maintenance.

3.   For large structures the metal is applied by spraying
and aluminium or zinc-aluminium alloys are an alternative
to zinc. Zinc dust is used in paints to give coatings with
some properties of a metal and some of a paint.

4.   For small parts, cadmium plating may be used as an
alternative to zinc plating, but plated coatings are only
economic when thin and hence only suited for mild
conditions or short lives.

APPLICATION OF ZINC AND ALUMINIUM COATINGS

5.   Zinc coatings can be applied by a range of techniques[1].
Sherardizing and hot dip galvanizing are diffusion-type
coatings in which the zinc coating is metallurgically
bonded to the steel by the iron-zinc alloys which form
during the coating operation. The nature and extent of the
alloys layers can be varied by the choice of steel, by
additions to the galvanizing bath and by variations in the
process. Electroplating and mechanical plating of zinc
and cadmium[2] rely on mechanical adhesion and - together
with sherardizing - are restricted mainly to small parts.

6. The most widely used process is hot dip galvanizing which is applicable to most forms of iron and steel and to both large and small parts. At the steelworks it can be applied to sheet and wire while they are still in continuous lengths. The steel is first cleaned, usually by oxidation/reduction techniques in continuous furnaces and then the strip or wire passes through molten zinc typically at 450°C. Addition of about 0.15% aluminium to the strip galvanizing bath restricts growth of the iron-zinc alloy layers and the resultant coating is fairly thin but extremely flexible and the coated sheet can be fabricated virtually to the same degree as the uncoated sheet. Galvanizing of products and of tube is normally carried out after semi-fabrication or fabrication. Cleaning is by cold hydrochloric or hot sulphuric acid pickling: the steel is then fluxed and dipped into molten zinc. Because bending and similar manipulation will mostly take place before coating, the iron-zinc alloy layers - which are at least as protective as the pure zinc layers - are allowed to grow naturally and the total coating is typically four times as thick as on continuously galvanized sheet; this means that the coating life is four times as long to first maintenance.

7. Both aluminium and zinc are applied by metal spraying. Essentially the basis metal is prepared by abrasive grit blasting using either compressed air or centrifugal blasting equipment. Chilled-iron grit is most widely used although steel grit, fused bauxite and other abrasives may be used.

8. The coating metal, introduced either as wire or powder, is melted in a stream of hot gas or an electric arc and projected as semi-molten droplets onto the metal surface. Metal spraying is, however, a 'cold' process, there being no significant heat input to the material being sprayed. Blasting and metal spraying can be done either in established works or on site, and in the case of large structures such as bridges it has proved economical to establish a special site-spraying shop.

9. For corrosion protection the thickness of zinc or aluminium applied should not be less than 100 μm nominal

(75 μm minimum) and thicker coatings are often both more desirable and more economical overall. Because of adhesion problems, gas sprayed aluminium should not normally be thicker than 250 μm, but this is more than sufficient to resist most atmospheres and other environments. Zinc can be applied more thickly and this can be useful because the life of a zinc coating in any given atmosphere is proportional to its thickness. Inspection of metal spraying is straightforward.[3]

SELECTION OF METAL COATING

10. The choice of coating method depends particularly on the size and shape of the article to be protected; the severity of the corrosive and abrasive conditions; joining requirements and initial long-term economics.

11. The combination of abrasion and corrosion resistance provided by galvanized steel makes it particularly suitable for uses where there may be some rough handling or where there is uncontrolled access for small boys with penknives. The choice of coating is often determined by the corrosion resistance required. For zinc this is roughly proportional to the thickness of the coating. In pure rural atmospheres in most countries zinc corrodes at 1-3 μm/year; in normal urban areas 3-6 μm/year and in marine atmospheres around 5 μm/year.[4] Corrosion in industrial atmospheres is heavier depending on the acid pollution. The purity of the zinc within the normal limits in coatings is not important but iron-zinc alloys, such as form part of the typical coating on galvanized products, have increased resistance to acid environments.

12. In the UK, information on coating lives and the selection of coatings is tabulated in the British Standards Institution documents DD24[5] and BS 5493[6]. Broadly speaking, only zinc or aluminium coatings of suitable thickness are recognised as being free of rust for more than 20 years in most atmospheres without maintenance.

13. Aesthetic or service considerations often make it desirable to apply paint or a plastic coating to the

galvanized material? There is a synergistic effect and the combined ('duplex') coating lasts longer than the sum of the lives of the two coating layers on their own. This is mainly due to the zinc layer preventing accelerated early failure of paint or plastic by the spread of rust under the coating from edges of any points of damage.

14. Fittings can be galvanized, sherardized or plated. Typically, galvanized fittings have about 45-65 μm zinc coating, sherardized 15-30 μm and plated 5-25 μm. For comparable protection coating thicknesses must be of the same order and although zinc-rich paints can be applied to add to the effective zinc thickness, it is preferable that the factory applied coating should provide the major protection against corrosion.

APPLICATIONS

15. Metal coated steelwork has been widely used in civil as well as structural engineering. Tunnels, culverts and bridges are well established uses of galvanized steel for both mass-produced components and for purpose-built structures. Metal spraying has been used for leading bridges worldwide, including the Forth, Severn and dozens of others in the UK. A lengthy and satisfactory case history of metal spraying is provided by the St Denis lock gates in France; galvanized bollards and other fittings are features of docks and canals. Trench sheeting is galvanized and bitumen coated to resist soil and water.

16. A special use of galvanized steel rod or mesh is in reinforced or prestressed concrete - good adhesion is obtained and the onset of rust (which can cause staining and then spalling of concrete from the internal pressures set up) is prevented or delayed. More recently, galvanized strip has been used in reinforced earth walls.

17. Some of these uses are discussed in the following sections together with other structural engineering uses which illustrate the durability and life of metal coated steel.

Culverts and tunnels

18.   Corrugated galvanized sheet sections can be bolted together to form tunnels large enough for rail or road traffic or smaller to provide a river or sewer passage. Freedom from maintenance is essential; the Swaziland railway, for example, was built in 1963/4 through regions with widely varying climates. Twelve years later, examinations showed all culverts to be structurally sound and only in isolated cases had any local damage or rusting occurred. No maintenance due to corrosion had been needed. In rapidly growing towns such as Caracas, the galvanized sheet tunnels provide a quick and effective means of constructing road interchanges.

Bridges and bridge components

19.   Multi-truss bridges have been widely used throughout the world for temporary structures or for access purposes. The Callender-Hamilton type has regularly been galvanized for the past 40 years. The coating ensures easy assembly and subsequent disassembly coupled with ability to withstand outdoor storage conditions which would cause rapid deterioration of steel without a zinc coating.

20.   The Westgate Bridge at Gloucester in the United Kingdom, erected in 1941, received no corrosion protection maintenance on the structure throughout its life; it was taken down in 1973 as part of a road improvement scheme. About six years before this one of the trusses was bent in collision but no structural weakness occurred and as the coating remained adherent and continued to protect the steel no maintenance was needed.

21.   The Callender-Hamilton bridges have friction grip joints with mating surfaces and bolts both galvanized. Extensive studies in the past 15 years have confirmed that hot dip galvanizing has suitable characteristics for high strength friction grip joints[9], but the required slip factor is only obtained after one or two cycles of slip under dynamic loading conditions. To avoid such initial slip it is necessary to roughen the galvanized surface or to apply a coat of zinc silicate paint to the faying surface

to ensure a high frictional coefficient. There is now 15 years experience of friction grip joints in galvanized arch and girder bridges also, starting with the Lizotte Bridge in Canada and now extending to about 100 bridges in North America; others are galvanized or zinc sprayed.

22. Bearing pads and expansion joints are also galvanized - a testimony to the load-bearing and abrasion resistant characteristics of galvanized steel.

23. Major river and road bridges often have components too large for galvanizing. So, when the Forth Bridge was built a metal spraying plant was set up nearby and the structural components were specified for zinc spraying and a four coat paint scheme. In practice, some parts missed the final paint scheme but were no worse for this when after 12 years some maintenance of the bridge structure was initiated - but less than 5% required attention other than for aesthetics. The bridge railings were galvanized and painted and - although the paint was damaged through the need to rattle the railings with a stick each day to dislodge seagulls - the steel remained protected. Some access covers were zinc sprayed without further treatment and were in good condition when examined after 12 years. The bridge cables were all galvanized and set into massive concrete anchorages at each end of the bridge.

Building components

24. In single storey structures and many engineering projects steel can be erected freely exposed to the surroundings and galvanizing can enable the construction to remain clean with no rust stains to affect nearby material or products. This is particularly important with wet process industries where there is frequently condensed water dripping from the steelwork. One typical warehouse for foodstuffs provides an 80m clear span through the use of galvanized steel arches. Galvanizing ensures that no rust in water drops will stain the packages stored beneath and force the sale of these as old or damaged goods. In a newsprint warehouse any rusty water droplets could materially damage the bales of

newsprint; the very large trusses in one building were welded after galvanizing because of lack of suitable large galvanizing facilities. Extensive research in the 1960's enabled positive recommendations to be made for welding zinc coated steel to give sound welds with no loss of strength[10]. This building, like many others, was given aesthetic appeal by cladding with painted galvanized steel sheets.

25. The galvanized structure itself can be used as an aesthetic feature as in the Whitbread Brewery building in Lancashire where the external mullions are left as-galvanized, contrasting attractively with the paint + galvanize coating on the infill panels.

26. Computer programmed storage facilities for goods in pallets are now being adopted in the United Kingdom. One warehouse at Purfleet, Essex, has 4,000 storage positions for pallets, each reached by cranes running in narrow channels between the storage areas. Close dimensional tolerances were necessary; this was met by galvanized steel - freedom from distortion during galvanizing was critical for this application and readily obtained.

27. Swimming pools too are particularly corrosive atmospheres due to the high humidity. The roof structure at Drancy, France, is made from tubes welded together after galvanizing with the weld joints touched up with zinc-rich paints. A more conventional roof structure is that of the pool at Düsseldorf Hilton Hotel which also shows the use of zinc rich paint on the joints of the air condition-ing ducts to restore protection to areas around welds where the zinc may have been volatilised. The success of galvanized steel for swimming bath structures is shown by the unrusted appearance of the French Boulogne-Billancourt pool roof structure after 10 years use. With cold stores, conditions can also be very corrosive and this has led to widespread use of galvanizing.

28. Space frames are very valuable for any area which has to be covered with a minimum of supports such as the Mildenhall freight terminal. This very large roof structure

was first assembled on the ground from galvanized
components, the individual joints being made by friction
grip connections, and then jacked into position. A more
specialised structure is the dome of the Imperial War
Museum in London which had to be rebuilt following a fire.
Here grit blasting preceded galvanizing to ensure a coating
nearly twice as thick as normal which would ensure
freedom from rust for the life of the building. Both
friction grip bolting (using galvanized faying surfaces and
galvanized bolts) and welding after galvanizing were used
in this structure.

29. Over 15 million tons of continuously galvanized steel
is produced each year in the world. This enables users to
fabricate a wide range of both utilitarian and decorative
components without further finishing. In particular,
cladding and roofing of buildings provides great scope for
this material and when necessary special design
techniques have been developed to facilitate detailing.
Unpainted sheet is widely used within buildings for decking.
Ducts of all types are prefabricated. Painting can also be
done by the user to suit his aesthetic taste as with garage
doors.

Chemical Engineering

30. Plants for many chemical products often have a
relatively short design life and this has in the past
militated against the use of high quality protection systems
such as galvanizing, but the increasing ability of
galvanizing to compete with paints on first cost has led
to more and more exterior steelwork in these plants being
zinc coated; aluminium or zinc spray is also used.

31. Recent UK examples include the Courtaulds synthetic
resin plants at Carrickfergus and Letterkenny, and the
Lennig organic chemicals plant in the highly polluted
area of Teesside - with galvanizing again specified for
further construction after evaluation of the performance
of the first plant erected nearly four years ago.

32. Where conditions are particularly corrosive,
galvanized steel is often coated with specially developed
paint systems. In the Hilton Davies plant for the

production of organic chemicals the galvanized steel is given a single coat of chlorinated rubber paint pigmented with zinc phosphate. One part of the structure was painted on-site but subsequent parts were painted in-works as it was a simple matter to touch-up any paint damage after erection - the galvanized coating underneath was an assurance against rusting. This enabled scaffolding to be removed more quickly than with on-site painting.

33.   The oil industry has tended to use zinc rich paints for many of their structures in the North Sea following successful use over many years in the Caribbean, but galvanizing has a substantial use for helicopter pads, communications towers, railings, stairways, flooring and other ancillary steelwork which requires resistance to the combined effects of corrosion and abrasion in the North Sea.

34.   On shore, galvanizing is extensively used both for the North Sea gas terminals such as that at Bacton, and for the oil terminals; the one being built for the Philips consortium on Teesside will utilise some 4, 500 tonnes of galvanized steel.

35.   Mechanical equipment in plants or for transport is required to withstand rugged treatment; the conveyor at the Immingham super terminal for coal and iron ore has a hot dip galvanized structure - the success of this led to specification of 8, 000 tonnes of galvanized steel for the Hunterston iron ore terminal. Many bucket conveyors used in quarrying or similar operations have both structures and buckets hot dip galvanized or metal sprayed.

36.   Concern at preservation of the environment has led to a rapid increase and modernisation of sewage treatment facilities. New sewage works use zinc coated steel for the rotating bridges and other exposed steelwork over treatment tanks; for immersed equipment such as scrapers and also for screws (typically with a malleable epoxy overcoat).

Reinforced concrete

37.   Where concrete construction is preferred to steel,

galvanized steel has an important role as reinforcement and research work has also confirmed its suitability in prestressed concrete. A minute quantity of chromate either on the wire or in the concrete (in many countries often a natural occurrence) is recommended. The use of galvanizing allows the depth of cover over the reinforcement to be reduced or, alternatively, gives greater assurance that the concrete will be free from spalling or staining in service.

38.   In the North Sea, the Andoc platform incorporates a sea-bed oil storage tank, from which the oil is taken off by tanker. The temperature of the oil, which helps set up stresses and cracks in the concrete, together with the corrosive effect of sea water on steel, led to this concrete tank roof being reinforced with 2,000 tonnes of galvanized steel.

39.   Galvanized reinforcement is particularly suited to docks where chloride contamination could lead to rusty steel and spalling of concrete. In Bermuda[12] where the coral aggregate used for making concrete used not to be washed because of shortage of fresh water, galvanized reinforcement is regularly used. This includes an 8 year old extension to No 1 Dock in Hamilton and the relatively new No 8 Dock in which the concrete part of the dock walls extends below low-tide level. An interesting additional use here of galvanized steel is the angle section used as a capping to the edge of the dock to absorb knocks and avoid chipping of the concrete. A fresh water tank which was demolished after 25 years showed the galvanized reinforcement still in good condition. NASA officials in Bermuda who have used both black and galvanized reinforcement for many years, take the view that one inch of cover is sufficient over galvanized reinforcement where two inches would be used over ungalvanized.

40.   Zinc coating of reinforcement for concrete adds to its direct cost but often the total cost of the project can be reduced because of the reduction possible in depth of cover required over the reinforcement. Maintenance

costs again accentuate the case for galvanized reinforcement. Bridge engineers in particular are greatly concerned at the cost of replacing bridge decks when these become unusable through corrosion of the underlying steel if this has not been adequately protected. As with many engineering projects, the price of safety or of improved performance is negligible when compared with the cost of occasional failures or unexpected short lives, or modifications and maintenance in service. Recently constructed bridge decks include the A6 Highway at Pont d'Ouche in France and 40 bridges on new elevated highways in Philadelphia, USA.

41. Building components frequently contain galvanized reinforcement to obtain slim lines and improve aesthetics.

Reinforced Earth

42. New techniques for construction of earth walls such as in highway cuttings enable walls to be more nearly perpendicular through incorporation of metal reinforcement. For the 50 year life normally specified, the steel is galvanized - tests are also made to ensure that the earth is not unduly corrosive. One of the first applications in the UK is at a road interchange near Poole.

43. An interesting use of galvanized wire and netting is on the surface of retaining walls to hold rocks and stones in position until natural bonding occurs.

COST OF PROTECTIVE COATINGS

44. In general, the initial cost of a metallic sprayed coating is no greater than the cost of a conventional high quality four or six coat paint scheme. With hot dip galvanizing the nature of the process gives a different cost structure so that galvanizing is equally competitive in first cost with painting for medium to heavy structural steelwork, while on the lighter sections it shows a substantial cost advantage. The exact cost relationships do, however, vary from country to country and with the distance from the nearest suitable galvanizing plant; metal spraying and painting can often be carried out on

or near the site.

45.   In many civil engineering projects there is little
scope for maintenance in service and in such cases the
extra protection against corrosion afforded by the zinc or
aluminium coating is well worth the small extra charge.
Elsewhere, the greatest economic advantage of metal
coatings is the savings they permit on maintenance costs,
which are continuing to escalate not only in absolute
values but also as a percentage of the total cost of a
project.

REFERENCES

1.   Technical Notes on Zinc - Zinc Coatings (A4 12pp)
Zinc Development Association, 34 Berkeley Square,
London W1X 6AJ
Also leaflets on the individual processes - galvanizing,
sherardizing, zinc plating, zinc spraying, zinc dust.

2.   Technical Notes on Cadmium - Cadmium Coatings
(A4 4pp) Cadmium Association, 34 Berkeley Square,
London W1X 6AJ

3.   (a) Inspection of Zinc Sprayed Coatings (A4 8pp)
          Zinc Development Association
     (b) Inspection of Sprayed Aluminium Coatings
          Association of Metal Sprayers, Chamber of
          Commerce House, Ward Street, Walsall WS1 2AG

4.   SCHIKORR, G Atmospheric Corrosion Resistance
of Zinc (A4 44pp)
Zinc Development Association

5.   DD24 - Methods of Protection against Corrosion of
Light Weight Section Steel used in Building (A4 30pp)
British Standards Institution, 2 Park Street,
London W1A 2BS

6. BS 5493 - Protective Coating of Iron and Steel
Structures against Corrosion (A4 110pp)
British Standards Institution

7. van EIJNSBERGEN, J F H   Twenty Years of
Duplex Systems - Galvanizing + Painting.
Paper to the Eleventh International Galvanizing
Conference, Madrid 1976. In conference volume
published by Industrial Newspapers Limited,
Queensway House, 2 Queensway, Redhill, Surrey,
RH1 1QS

8. Galvanizing for Structural Steelwork (A4 20pp)
Galvanizers Association, 34 Berkeley Square,
London W1X 6AJ

9. MOORE, R et al   Galvanized Steel in Friction Grip
Connections. Construction Steelwork Metals and
Materials 1970 (December). Reprint available from
ZDA

10. Welding Zinc Coated Steel (A5 131pp) American
Welding Society

11. PORTER, F C Comparative Costs of Protecting
Steel (A4 9pp) 1978. ZDA/LDA

12. PORTER, F C   Galvanizing of Steel Reinforcement.
Concrete 1976 (August) (A4 4pp) Reprint available
from ZDA.

10 Quality control of protective coatings

D. A. BAYLISS, FICorrT, FTSC, Technical Director,
BIE Anti-Corrosion Ltd

SYNOPSIS. Unlike many other engineering processes in the
construction industry, for example welding or concrete
production, the failure of a paint or coating system on
steelwork seldom results in a dangerous, structural
failure, yet there is a growing awareness that modern
coatings and their application cost a great deal of money
and too often do not live up to their expectations. It
is also unfortunate that many modern materials seldom grow
old gracefully but either last for a very long time or
fail quickly on a large and dramatic scale. In addition,
so many modern protective systems applied initially at
works cannot be replaced to the same standard insitu and
often the user is left with a legacy of maintenance
painting with progressively shorter intervals between
repaints.

It does not help the situation that the final protective
film is the product of two separate entities - the paint
manufacturer and the paint applicator. Also the final
product itself - to quote BS5493 Code of Practice for
"Protective Coating of Iron and Steel structures against
Corrosion" - "is susceptible to operator abuse or adverse
environmental influences throughout all stages of the
work. Furthermore it is generally difficult to deduce
from examination of the completed work what has occurred.
Consequently, the coating may fail prematurely but more
often the effect is a reduction in long term durability".
In other words paint is very good, at least on a temporary
scale, in covering up transgressions.

For many engineers corrosion protection is a minor part
of their work and more often than not, because of the
multitude of factors that can be involved, is a nuisance

out of all proportion to the end effect. Yet for cosmetic or for health and safety reasons there are very few structures that can be left unpainted.

It follows that for any painting work of consequence for example if maintenance is difficult or extremely costly, there must be full-time monitoring of the process. Ideally this should be carried out by a member of the customers staff working solely under his control and solely with his interest at heart. However often even if there is such a person suitably qualified for such work he seldom can be spared on a full time basis; in other words to be with the applicators for the entire time that they are working. In such cases this inevitably means that when his back is turned areas will be prepared or painted at a faster rate or in a less meticulous manner in order to make up for the loss of bonus whilst he was watching. The areas are often indistinguishable from the remainder but are likely to show localised breakdown within an appreciably shorter period.

Nowadays most Engineers concerned with industrial painting are aware of the existence of independent specialist inspection firms. Many have used their services, some very successfully, some less so. One cause for the latter is that some inspectors are casual employment from agencies or from organisations that have inadequate back-up expertise and training facilities. A good painting inspector should know 'why' as well as 'how'. The object of this paper is to highlight some of the factors that he and his customers should be aware of during the coating process.

Surface Preparation.

Most people appreciate that surface preparation is a vital factor in coating durability. Unfortunately many pay lip service to this requirement by a bald statement in a specification that the surface will be blast cleaned to SA3. Freedom from residual millscale is certainly a factor to be considered but not the only one. The following are considered to be the more important properties required of a surface to be coated.

a) The surface must be firm and free from any layers that will prevent the paint film wetting the surface and adhering, for example loose flaking paint, dust, grease, water or even loose 'hackles' or metal caused by the abrasive cleaning process.

b) A surface must be chemically clean, in particular, free from corrosion producing matter such as ferrous corrosion salts, weld deposits etc.

c) The surface must be of the right roughness both from a point of view of minimum profile and maximum profile for the coating to be applied. None of the above factors are identified by use of the Swedish Standard photographs. The Swedish Standard also refers solely to previously unpainted steelwork whereas the majority of painting work is carried out for maintenance purposes.

Another important feature not covered by the photographs or the text although it is referred to in BS4232 (a more complete document but much less widely used than the Swedish Standard and now under review), is the roughness of the surface. It is obvious that the surface must not be too rough in relation to the thickness of the coating. It is less obvious and less widely specified that for coatings with high cohesion, the surface needs to be of a sufficient roughness to prevent loss of adhesion by peeling. Modern paint films will adhere remarkably well to smooth surfaces, for example glass but particularly with high build epoxy coatings their cohesive strength is such that leverage under the film can easily strip a coating from a smooth surface. The effect can be seen in poor resistance to mechanical damage. In terms of the tests that can be carried out on the film, this means that the pull-off adhesion can be quite high whereas the peeling adhesion is very poor. If the surface is slightly roughened then it appears that the protuberances on the surface can produce a notch effect in the film and limit the extent of the damage.

To specify for high build coatings that the maximum blast profiles should not exceed 25% of the total film thickness and that the minimum blast profile should be not less than 35 microns, would seem to be a satisfactory and practical compromise. However it is also necessary to take into consideration the limitations of the methods of profile measurement. Work carried out at P.R.A. in connection with the proposed new ISO standard on surface preparation for painting looked at ten different types of profile measuring equipment. The results gave a confused picture with very poor agreement between types and with the reference method.

CORROSION IN CIVIL ENGINEERING

The lesson therefore for quality control is that no profile measurement should be specified or attempted to be controlled to very exact limits. Apart from careful choice of abrasive size, what appears to be needed is a rapid method to give numerical values which can be correlated with good practice.

The problems of painting rusty, pitted steel and of welds are also, in part, due to the surface profile on a rusted surface. There is no difficulty in finding peak to trough heights generally well in excess of those obtained with coarse abrasive cleaning.

As previously mentioned another important feature of preparing a surface to receive a coating is that the surface must be free from dust and debris. The test accompanying the photographs to the Swedish Standard states that all foreign matter must be removed and the surface vacuum cleaned. This text is often ignored in practice but in any case for the purpose of quality control it is not specific enough. What is foreign matter? What methods of test can you use to ensure that your cleaning methods have been adequate? For example there is no standard method to determine the presence of a fine layer of dust on the surface and yet this is often difficult to see by eye and can cause very poor adhesion for subsequent sprayed applied coatings. An adhesive tape test can be useful but the inspector needs guidance as to what percentage dirt he can accept on the test tape for any specific material.

Similarly a thin film of moisture can be equally difficult to detect and equally disruptive. Such a film can occur when a large metal object is brought from a cold storage area into a warm shop to be painted. In these cases a suitable moisture indicating test paper would be useful and certainly preferable to somebody feeling the surface with their bare hands!

A thin layer of oil or grease on a surface will also prevent it being wetted by subsequent paint films. Specifications talk loosely about the removal of oil and grease but often without specifying how this can be achieved and how it can be determined that it has been achieved. Solvent degreasing in particular, as opposed to the use of emulsifiable degreasers, can deposit a thin film of oil or grease over the entire surface. Such a film is also difficult to detect by eye and it is preferable to use a test method such as the water break test or the Fettrot dye test.

The test, which is given in DIN Standard 55 928, consists of applying to the surface a drop of 0.1% solution of the dye Fettrot BB ethanol (the test will work equally well with 1% solution of crystal violet or fluorescein in ethanol). On a horizontal grease-free surface, the drop quickly spreads, leaving a circular rim. On a vertical grease-free surface, the path of the drop is short and an oval rim remains. On a horizontal surface which is not free from grease, the drop keeps its initial size and after evaporation a sharply serrated edge is apparent. On a vertical surface which is not free from grease a long trail is formed.

Contamination can also occur on blast cleaned surfaces prior to painting due to local pollution or sea spray etc. Salt crystals can be seen on a dry surface but not a wet surface therefore it is sometimes necessary to carry out a chemical test.

This can be carried out by soaking a filter paper in silver nitrate solution (20 grams per litre) and allowing it to drain but not dry. The filter paper is then pressed onto the surface using plastic clay to ensure intimate contact. After 20 seconds contact the paper is removed and washed thoroughly in distilled water, using not less than 3 separate washings. The paper is then immersed in normal photographic paper developer for two minutes. Chlorides will show up as brown black spots. The paper may then be washed and dried and kept as a permanent record.

An even simpler qualitative test can be used when the surface is wet after washing down. This is to scrub a standard area with a painting pad (a polyurethane foam pad with a bristle surface backed with a metal handle). The pad is then immersed several times in distilled water after which silver nitrate solution is added and the turbidity produced visually compared with a standard sample containing the washing water. The surface could be considered reasonably free of chloride if the test solution was no more turbid than the washing water solution.

Another form of contamination which can be invisible on a
white metal blast cleaned finish is that of corrosion
producing ferrous salts. These will generally be present
on any previously rusted surface and are of particular
significance where the painted surface will be under
immersed or condensing conditions. The ferricyanide test
as described in BS5493 appendix G is of value in indicating
the presence of such salts. It is however a difficult test
for an inspector to interpret. For example, is any degree
of blue colouration tolerated? Quantitative methods
indicating the amount of ferrous salts over a sample test
area will eventually be more valuable when the quantities
can be related to good practice.

The low powered magnifying glass, for example x 15 is, as
usual, a particularly valuable inspection tool for surface
preparation. At such a magnification the nearly colourless
crystals of ferrous sulphate or ferrous chloride can just
be seen at the bottom of the pits of the white metal
cleaned surface. It should be noted that higher magnifi-
cations are generally less valuable as the higher intensity
of light required makes the crystals less easy to distin-
guish.

Another undesirable presence on a surface to be painted
is that of laminar millscale. If present in sufficient
quantities it is likely to become detached by corrosion
or thermal expansion and cause flaking of the paint film.
Swedish pictorial standards SA3 SA2. etc are generally used
to detect the presence of residual millscale on new steel,
Unfortunately these standards were derived only from dry
blasting techniques and since it may well be more effective
to remove the major part of contamination as mentioned
above by the incorporation of water, a new set of photographic standards is required. It may well be that a less
than white finish but one more chemically clean would be
a more satisfactory surface for painting.

In addition by whatever method a surface is prepared it is
essential to inspect and rectify any defects in that surface
that have been revealed, for example, skin laminations,
surface shelling, rolling laps, cracking and deep pitting.
Such blemishes can seldom be adequately coated.

A further vital and again frequently neglected aspect of
surface preparation is the treatment of weld areas. The
neglect is possibly due to the fact that it represents
only a small proportion of the total area and yet in

practice is the commonest point of breakdown to occur and,
in view of the possibility of stress corrosion, the most
undesirable. The problem is two fold, firstly the weld
itself can consist of very sharp points, cavities, blow
holes etc which are almost impossible to paint and
secondly the alkaline contamination that occurs from the
welding process. For this latter reason weld areas should
be checked with pH papers.

PAINT APPLICATION.

Unauthorised thinning of paints, particularly to make high
build materials easier to apply, is the commonest grounds
for fault at this stage. Quality control testing consists
of weight per gallon determinations on samples from a
painter's kettle. These can be carried out on site,
particularly in comparison with fresh samples from the
paint tin, but often the taking of samples alone is
normally a sufficient deterrent.

Other major faults are the inadequate stirring of heavily
pigmented materials, such as, micaceous iron oxide paints
or zinc rich primers, or the incorrect mixing of two
component materials. Another less easily detected fault
but which nevertheless can cause subsequent severe loss
of adhesion, is the use of two component mixes when they
appear to be satisfactory but have in fact exceeded their
pot life. Such partially cured materials will generally
not 'wet' the surface to which they are applied.

INSPECTION OF APPLICATION OF PAINT.

Techniques for the measurement of dry film thickness of
paint are well known. However it should be remembered
that for a variety of reasons they have limited accuracy.
For example the manufacturers of the magnetic type gauges
only claim an accuracy of plus or minus three percent for
the most accurate of their instruments and anything up to
plus or minus ten percent for others. In particular,
variations in surface roughness are of even greater
significance even when the instrument has been calibrated
on part of such surface. Measurements taken on a completed
coating without calibration on the original surface and
without knowledge of the state of the surface coated can
only be a rough guide. Inadequate cleaning and a very
rough surface both contribute to the film thickness
indicated. Wet film thickness measurements taken during
application are valuable since they allow instant

CORROSION IN CIVIL ENGINEERING

correction and this is a typical example where quality
control inspection can help rather than hinder the
progress of work. Contrasting colours between coats is a
simple but useful quality control technique and conversely,
single coat high build materials are both difficult to
spray uniformly and to avoid dry spray areas. Holiday
detection of the correct type and at the correct voltage
should be carried out on coatings which are to be used
in submerged, corrosive conditions.

AMBIENT CONDITIONS

An inspector needs to keep a constant eye on air and
surface temperatures, humidity, rain etc. The majority
of specifications state that paint shall not be applied
or blasting shall not take place when any of the following
conditions exist (a) the metal temperature is less than
5 degrees F above dew point (b) the air temperature is
below $40^{O}F$ (c) the relative humidity exceeds 85%. In
fact for some materials this is too restrictive, for
example inorganic zinc silicates and others not restrictive
enough, for example two pack polyurethanes. The main
criteria for inspection should be obtained from the
manufacturers data sheets for the product and manufacturers
should overcome their natural reluctance and indicate
under what conditions their materials are less than
perfect. It would also help users considerably if such
information from all manufacturers was presented in a more
standardised form.

RECORDS

The keeping of clear and accurate records is an essential
part of quality control. In practice there is a difficult
line to draw between obtaining insufficient information
and generating so much paperwork that the headquarters
scrutineering and summary reporting systems are overwhelmed
Two important aspects of reporting are that areas reported
on must be clearly and exactly identified and that records
should be retained for possible reference in the future.
Retention should be a minimum of five years for conventional
paint systems and appropriately longer for high performance
coatings. Unless an inspection company transfers this
information to micro-film there can be an embarrassment
for storage and for possible retrieval, at a later date.

AIMS AND DUTIES OF A GOOD QUALITY CONTROL INSPECTOR.

Since inspection work can cover a number of different

materials such as conventional paints, metal coatings, powder coatings, hot applied solventless epoxies, fusion bonded coatings, it is essential that the inspector has sufficient training to meet the different requirements of different materials. It is unlikely, in fact probably undesirable, that he will have a very high standard of academic education or theoretical knowledge. In fact if he has, the customer would probably be unable to afford his services. That does mean that the majority of quality control inspectors in the field should not be used for solving other painting problems, failures etc. or in providing information for writing specifications etc.

It is easy to draw up a list of the qualities required of the ideal painting inspector but again one must take a practical attitude. The job is not the most congenial, for instance including long hours and long periods away from home. The work like painting itself can be unpleasant and even with an element of danger. In practice the difference between a good and troublesome job is often a measure of the inspectors' personality. A bad inspector is like a bad policeman and is either too petty and rigid or the reverse. Good quality control should result in the satisfaction of all parties that everybody has done the best they can, to the best of their abilities.

SELECTION OF MATERIALS.

It is an interesting fact that whereas it would seem that the first consideration in selecting protective coating would be the environment to which it has to withstand, this is only partially true. The type and standard of surface preparation possible for any particular situation often has a far greater influence. Where for example abrasive cleaning is either impractical or unlikely to be carried out to a high standard, it is normal to use less durable but more tolerant materials such as for example bitumens alternatively one progresses up the scale in, less tolerance and more durability, to BS2523 red lead, chlorinated rubbers, pitch epoxies, solvent free epoxies for example.

Quality control of paint constituents can be a valuable exercise particularly if materials with possible variations in raw materials are used. However by far the greatest value can be achieved by the quality control of painting so that the coating with the right standard of durability can be used in the right situation.

Discussion on Papers 1, 2 and 3

MR GOODMAN

Reference in Paper 2 to MIO is to conventional or oleoresinous MIO.

I am indebted to the East Sussex County Council for their permission to examine the River Iden Bridge which was constructed in weathering steel.

In relation to the difficulty of achieving effective maintenance, I suggest that those without direct experience of the problem should ask themselves how often they could be sure of obtaining the necessary conditions which I define as possession of the structure for a sufficient period of time in suitable environment conditions to carry out the work necessary by the methods appropriate to the condition of the surface or substrate. To talk of long life is meaningless unless effective maintenance can be carried out when required.

MR D. NORMAN, British Gas Corporation

I agree with Mr Goodman that it is important to be realistic. Too many corrosion oriented people tend to be unrealistic in regard to coating use and coating provision.

I agree that a new coating system should be chosen with subsequent maintenance in mind. This is all too rarely done. In deciding which treatment is best suited to a particular type of coating work the features listed in paragraph 6 of Paper 2 should all be taken into account.

The system chosen must be the best possible as there can be no guarantee that maintenance will be carried out when it is first needed.

How does one ensure that coatings applied in a factory environment and in a works environment are quality controlled?

CORROSION IN CIVIL ENGINEERING

MR A.N. McKELVIE, Paint Research Association

Damage on epoxy paint from missiles is a problem in the
motorcar industry, where there is chipping along the bottom
of cars. Volvo have solved this problem by using an inter-
mediate coating which absorbs the energy and thus eliminates
chipping. Has British Rail considered this approach?

It need not take years of research to test a solution to
this problem. A quick result will be found by painting on
the material and throwing stones at it. If it does not come
off it will be alright.

MR N.S. MAKINS, Constructional Steel Research and Development
Organisation

In general it is fairly easy to protect the exposed surfaces
against corrosion. The problem areas are more often at the
ends of members, crevices and joints. Is it feasible to
have preferential protective systems at these problem areas?
Do British Railways carry out selective protection where it
is known that corrosion will be a problem?

Do they do anything about fretting corrosion, or is it no
longer a problem?

DR R. JONES, University College, Cardiff

In paragraph 11 of Paper 2 Mr Goodman gives details of an
advisory group. Was it deliberate that a metallurgist was
not included?

MR W.E. WAKEFIELD, Unilever Ltd

How long did it take the advisory group mentioned in Paper 2
to come to its initial conclusion and recommendations?
Presumably the group considers advances in technology. How
does it present its findings to British Railways for internal use?

MR I.P. HAIGH, Sir Alexander Gibb & Partners

CP 2008[1] was drafted in the Institution of Civil Engineers by
a committee under the chairmanship of Dr J.C. Hudson and
serviced by A.R. Jesty. Despite Mr Chandler's criticism of
it, CP 2008 was an excellent guide to practice, and its
replacement only twelve years later by BS 5493 is an indication of the rapid developments in the science and technology
of corrosion protection and in the research effort made

within the construction industry itself. This research is now in the hands of CIRIA which collaborates well with other, more senior, research associations and which earned the particular commendation of the Bessborough Committee for resisting the temptation to set up its own laboratory; CIRIA commissions work in other laboratories, such as the Paint Research Association which has the necessary skills and facilities.

One of the best features of BS 5493 is its structural form. It cannot be used as a casual reference; it is necessary to read it in a logical fashion in order to select the optimum painting system which, by still following the advice given in the code, hopefully can then be properly implemented.

Nevertheless any code of practice suffers because it has to be a consensus document. All interests must be taken into account without showing bias. This requirement tends to make codes rather indecisive - a weakness illustrated by the number of systems tabulated in section 2 of the new code.

Codes should not be textbooks. They should give advice on what should be done, how it should be done and on the results to be achieved. They should not discuss the reasons for giving this advice; that is the role for supplementary documents which are often written by people who take part in drafting the codes.

In the case of BS 5493 two such efforts are being made, both of them involving Mr Chandler. One is the series of guides to practice which is being prepared on a range of topics by the Committee on Corrosion of the Department of Industry; the other, restricted to the narrower, but important, subject of protecting steelwork by painting, is to be the work of CIRIA.

The opinion that British Standard codes should not attempt to discuss the reasons for, or to justify, the advice given does not imply that engineers need not endeavour to understand the corrosion processes, to discover how corrosion can be countered, and to find out how protective coatings work and what is in them.

However, few engineers make this effort. They regard coating materials as complex organic and inorganic chemicals, beyond their knowledge of chemistry. This is a mistake; they should be as familiar with anti-corrosion coatings as they are with soils, concrete or asphalt.

Indeed, there are close parallels between paints and these other materials in that all of them are composites, consisting of particles held in a matrix of another material. The constituents of paint, known by other names, are no more

than the aggregates, cement, admixtures and water to be found
in concrete. With each of these materials, the aims are to
produce a composite that is sufficiently strong, ductile,
chemically stable, impervious and weather resistant, and as
free of voids, cracks and capillaries as is practicable.

Until engineers know as much about paints as they do about
concrete, they should seek advice from independent experts on
corrosion protection at an early stage in the project, long
before the difficulties arise.

Modern paints incorporating synthetic media require much
more thorough preparation, more favourable conditions for
painting and more careful application than traditional paints
based on linseed oil.

At the present stage of developing practices, only a
method specification will serve; one cannot rely on a performance specification or on a guarantee.

The paint specification is not only a contract document;
it also provides the engineer with the opportunity of thinking out exactly what he seeks to achieve and how he will
achieve it.

Paints applied to real structures will not necessarily
perform as well in service as the same paints applied expert-
ly to small coupon samples in a laboratory and exposed to the
same climate.

An important step in selecting the painting system is to
determine how much of that system will be applied in the shops,
before the work is transported to the site, and how much will
be applied after the steelwork has been erected.

It is more difficult to protect the edges of rolled sections, the welds and mechanical joints in fabricated steelwork than the plain webs and flanges.

A less apologetic and defensive attitude should be adopted towards painting steelwork; unlike some other materials,
such as concrete, steel surfaces accept paint well, and there
is no better way of improving the appearance of a dirty and
old structure than by repainting it.

MR R. ALCOCK, British Railways, R & D Division

In view of their need to have long-life protective systems
which require minimum painting on site, British Railways
favour a metal spray coating plus paint coating for bridges.
These can be largely applied in a fabricator's workshops.
For the very large pieces of steelwork involved there is no
doubt that metal coating plus paint gives the fewest problems when it comes to maintenance, particularly from an abrasion and impact point of view.

Newer types of coating potentially have a long life, but if something goes wrong, often it is in an incipient way which does not reveal itself until possibly it is too late to rectify, because when maintenance painting is deemed to be necessary it can take up to five years to make money available.

MR F.D. TIMMINS, Mebon Paints Ltd

Red lead pigments do not facilitate modern formulation techniques such as viscosity and sag control. Further, they present unacceptable health hazards at all stages.

It is claimed that certain British Rail structures bear up to 100 tons of lead based paint. Ultimate elimination of this hazard would appear to be virtually impossible.

MR R.H. CUTTS, Property Services Agency

CP 2008[1] put over the concept of surface preparation; I think BS 5493 puts over the concept of thickness of coatings. Eleven years elapsed between the publication of these codes. I wonder if it will take as long for coatings applied in controlled workshop conditions to be accepted as the norm. I am surprised that this did not receive greater emphasis in BS 5493 because the one time when steel is completely accessible is between fabrication and despatch to site and materials being used are very sensitive to ambient conditions. Inspection at shop floor level is easier than on site. Coatings applied in the shop will have their problems but an associated problem which requires particular attention is the enhanced protection of fasteners.

MR E.J. PEARSON, Taylor Woodrow International Ltd

One area on structural steelwork where there is a problem is on the corners and returns on fabricated details. These are often possible to design out, but there are always the sharp corners on the members themselves. These corners are often susceptible to corrosion because they are in the places where drips form. They are also susceptible to chipping, and they are areas where the coatings are most tested. Do British Railways specify stripe-painting or preferential treatment of those areas? How do they quantify the thickness of paint to be put on those areas?

When building structures, do British Railways provide brackets or rails for moving gantries so that access systems can be placed on them afterwards?

CORROSION IN CIVIL ENGINEERING

MR E.N. CAPPITT, Posford, Pavry & Partners

I have had experience of studying codes of practice, talking
to paint manufacturers, technicians, special suppliers and
users and then recommending to a client that a particular
new technique is worth considering, only for the client to
reply that he has his own maintenance people who do not have
experience of the specialist techniques required to maintain
these systems which may need to be applied in factory controlled conditions by a specialist paint crew with special
experience.

How do techniques required for applying and repairing new
painting systems compare with well-known and tried systems?

MR W. SHEPHERD, Corrosion Protection Services

Steelwork fabricated in the UK which is going abroad is given
inadequate temporary protection to provide prevention from
damage in transit.

Thin sheet hot dip galvanized steel, usually in coil form
for making roof decks, wall cladding and so on, is particularly prone to a form of corrosion called white rust. In
severe cases this can lead to red rusting of the underlying
steel and has caused rejections in many cases. BS 5493 gives
no guidance to purchasers from abroad or suppliers from the
UK to prevent this. Orders are being placed in other
countries because of the poor history of rejection of British
galvanized strip due to white rust.

Fabricated structures for export are often given just an
etch primer or a red oxide primer and a lick of something
grey. They look fine when they leave the fabricator's shop,
but when they reach site they are useless. BS 5493 gives
little guidance on re-coating structures which cannot subse-
quently be blast cleaned.

MR McKELVIE

There is no doubt that if all the jobs could be done in the
fabrication shop this would give a much better end result.

With regard to quality assurance, government departments
ensure that the main contractor has the facilities to do the
job. Mainly he has the facilities to do the construction
job but he probably subcontracts the painting and protection.
I believe that until the protection is more closely integrated with production there will always be the risk of
failures.

MR A.G. McCONNELL, Hempels Industrial Coatings

For some years paint manufacturers have been stressing to fabrication engineers and specifiers the importance of applying coatings in the workshop. The main resistance to this practice comes from steel fabricators who are looking for a quick through-put for their paint shops. In most cases they object to applying second and third coats in the workshop. I think engineers should be asked to make it clear in their tender documents and specifications that second, and possibly third, coats are to be applied in the workshop.

MR K.E.T. FORSYTH, Berger Protecton Special Coatings

It should be stressed that the characteristics of the paint coatings chosen in BS 5493 provide no more than a broad indication of the type of paint. They cannot be considered as being a formulation of paint composition. Such information can be misleading if used incorrectly and many coatings outside the parameters given in BS 5493 have successfully stood the test of time.

Many paint companies specialize in steel protection, and I think their advice should be sought at the design stage. They can offer specifications that are practical to the environment and to the user. Whether two or four coats of paint should be applied in the fabricator's workshop must not only rest on the question of steel protection but also consider how practical it is to apply four coats in the workshop. The paint manufacturer knowledgeable in steel protection can offer the most practical system to use.

DR CLARKE

Mr Norman asked about quality control of coatings, which is a matter in which I find great difficulties. One wants to know whether corrosion control systems work, but all direct tests of corrosion resistance are destructive. Therefore, one tests a very small sample. With a small sample the uncertainty of applying the results to the whole batch is great. Additional information is sought by testing for qualities believed to be related to corrosion. For example, a paint coating to protect a bridge from corrosion for 25 years is difficult to test directly. However, it is believed that it will help if it is the correct thickness. If it is not the right thickness, there are suspicions. Thickness may be measured non-destructively; hence a larger sample may be taken. Most specifications are drawn up on this approach. Rather than making a direct test, related qualities are measured, such as paint, pigment concentrations, thicknesses and pre-treatment. Even so, I find that there is difficulty in carrying out the specifications laid down for sampling. The cardinal rule is that the sample should be random. However, there is nothing random about a random sample, and it takes care and directed effort to get one. The manufacturer's welcome is equivocal because I am testing on behalf of the buyer. The batch is stacked in 100 ton loads to bury those parts underneath, and there are no fork-lift trucks available. The shed is not heated, and although there is no specification for testing below 10^0C, I have to test at 0^0C. Suppliers may be switched, because of a lower quotation, to Holland, Germany or somewhere inaccessible. When the batch is landed, passed through customs and found defective, the cry goes up: 'It cannot be sent back all that way. What can be done about it?' Great pressure arises to accept sub-standard material, because of the rush. This applies also to new work. Because of penalty clauses in the contract, the builders say, 'We want this up by the end of the year. We shall have to accept this sub-standard material.' Under these circumstances quality control is almost dead. However, if it is known that there are going to be tests, the consciences of some of the fabricators and applicators are touched a little and they will try to do better.

Over the years in corrosion engineers have commented, 'The systems whose failures you are showing are 20 years old. We do better now.' But how can one know that one is doing better until 20 years from now? After 20 years, is perfor- mance superior? No. But, of course, in 20 years' time one has even better systems!

I have suggested in my Paper an approach to practical corrosion like that used in medicine. One has to 'walk the wards' and look at corroding structures to see what the prob- lems are. One can retire to the laboratory and see whether one can solve them with existing knowledge, or whether one needs new knowledge. But I think it is true that many corro- sion enthusiasts - and I will admit to being one - retired to their laboratories years ago and have not come out. They are happy solving artificial corrosion problems. Because of the time factor, we always appear to be fighting the last war. This is true of anything one has to learn from history. Unless there is continuity in the systems we learn about, what use can we make of our lessons?

MR GOODMAN

In reply to Mr Norman, in my opinion the only way to ensure

quality is to create the right environment. This costs money
and, even so, is not always practicable. If a high standard
is essential then the conditions must be created in which it
can be achieved. If the conditions cannot be created then a
process must be used which does not require such a high
standard. I think it must be accepted that there is a point
where there is no benefit in doing anything. Supervision is
obviously necessary but is not much use if the physical
conditions are inadequate.

With regard to Mr McKelvie's comments, British Railways
do know about cushion coats. However, these coats have other
problems so far as systems for structural steel are
concerned.

As an engineer, I am not convinced that laboratory results
always indicate clearly what one can expect in practice.
They may give clear answers to one or two problems, but on site
there will be three problems. So because British Rail
needed solutions that were applicable to large numbers of
structures, they looked at all the structures they could find,
on and off the railway.

In reply to Mr Makins, I think it should be feasible in
many cases to provide preferential protective systems in
problem areas. To a limited extent this is done already.
Further development is desirable.

The design of structures has changed markedly in recent
years. I am not aware that fretting corrosion is a problem
in the more recently erected structures. More recent structures tend to be largely welded with high strength friction
connections at the joints which have to be connected
separately on site.

Dr Jones and Mr Wakefield asked about the advisory group.
It was deliberate that a metallurgist was not included
because if a group becomes too large it becomes ineffective.
British Rail employ many specialists who can be approached as
necessary. Because of the number of activities being
carried out at the same time there was probably about a year
between the group's decision and conclusion. However, the
group was able to draw on the results of earlier work so
this is not total time. The group develop recommendations
based on their findings. These range from simple recommendations to the production of handbooks and technical notes.
The group reports to a committee of senior railway engineers
dealing with works matters. When recommendations are
accepted they become departmental policy.

In reply to Mr Pearson, British Railways specify stripe-
painting in clause 5.7 of their general specification for
steelwork.[2] Provision is made for access but not neces-

sarily with fixed devices as use is also made of mobile
inspection gantries.

Mr Cappitt asked about the techniques used with new as
opposed to well-established painting systems. In fact they
are basically the same.

The crux of the matter is that one must think of what one
wants from a structure. It is not resistance to corrosion
by itself that one is really after, but a structure which
does what it is required to do for the period of time for
which it is required to do it. All activities must be
directed to this end. In this anti-corrosion work has a part
to play, but it must be directed at those areas where such
work matters.

MR CHANDLER

With regard to Mr Forsyth's last comment, the code does
recommend users to seek advice from the producers of coating
materials. To some extent the purpose of the code is to
provide the engineering or steel user with the means of
starting a meaningful discussion with the paint company.
There are many paint companies in the UK. Some have high
quality research and quality control departments, but others
have not. Clearly, it is in the interests of the paint
industry to make sure that people understand the nature and
problems of protective coatings so that they will be
encouraged to deal with the reputable companies.

In principle I agree that it is advantageous to coat steel
in the shop although there can be problems. If all but the
last coat is applied in the shop it means the steelwork must
be handled more carefully. Although this is desirable, one
must accept that practice and ideals are not always the same
thing. Nevertheless, paint applied under good conditions is
likely to provide the best performance.

Mr Shepherd made a number of points. This particular
standard does not deal with thin gauge material. There is a
draft for development (DD 24) which deals with thin gauge
steel. However, there are chromate treatments that can be
applied to prevent white rusting of galvanized coil.

With regard to the cleaning of steelwork, there is a
section on maintenance in BS 5493. Of course, if steel is
inadequately cleaned and painted and arrives overseas in a
poor condition, there is no effective way of treating it
other than by blast cleaning.

I think Mr Cappitt raised an important point. One has to
beware of making the protection of steel too sophisticated.
One must take into account the probability of success with the

more sophisticated coating systems. In many situations the standard of labour is such that a more easily applied, less protective system could be preferable. Sometimes coatings have to be maintained by staff who are not always of the highest calibre, often under conditions that are by no means ideal, and this must be taken into account.

Costs are obviously very important. Money is not always available to provide the degree of protection required but if that situation arises there are various ways of saving money. For example, if a blast-cleaned steel with a four coat system is required and this is too expensive, then rather than cleaning by wire brushing then applying the same four coat system, it may be preferable to consider a reduction in the number of paint coats but retaining the blast cleaning of the steel. The thickness can often be increased later.

With regard to Mr Haigh's comments, I did not mean to give the impression of being critical of CP 2008. I was involved in its preparation and consider that it represented an important breakthrough. By their nature, codes of practice rarely satisfy everybody but they should result in an overall improvement in practice and I hope that BS 5493 will at least achieve this.

References

1. British Standards Institution. Protection of iron and steel structures from corrosion. BSI, London, 1976, CP 2008.

2. British Railways. The general specification for steel-work. Civil Engineering Department, BR, London, 1979, Handbook 27.

Discussion on Papers 4 and 5

MR A.N. McKELVIE, Paint Research Association

The paint system concept is only the first step in the protection of new structures towards an ultimate goal of inte-
grating the protection function into the production process.
More careful consideration is required at the design stage,
and the right time and place for every stage of the protection process should be charted from the cardinal date plan
right down to the day to day production planning. What
progress has been made in the USA in the 1970s towards inte-
grating protection with production?

Blast cleaning is recognized as the best surface preparation for steel, but frequently there is failure where the
blast cleaning is said to have been satisfactory. I believe
the reason for this is the inadequacy of existing standards.
Too much reliance is placed on the Swedish pictorial standard.[1]
This has outlived its usefulness as the sole criterion for
judging blast cleaning quality, especially where the cleaning
of rusty steel is involved, and must be supplemented by
physical ahd chemical tests. The physical tests must charac-
terize the blast profile more precisely as this has a definite
bearing on the expected performance of the coating. Chemical
tests must identify any remaining rust producing products
which, although not visible when compared with the Swedish
standard, may be there in such quantity as to cause rapid
breakdown of the coating by blistering and rusting through.
Residue of iron oxide or rust is less of a problem than are
colourless rust-producing products left on so-called cleaned
steel. The potassium ferricyanide test detects colourless
ferrous salts remaining after blast cleaning and is used more
appropriately after dry blast cleaning as water abrasive
blasting is more likely to have removed these contaminants.

Is there a plan to introduce such testing into SP 5, SP 10,

and SP 6? Should not the brush-off grade SP 7 be made obso-
lete? Should it not be recognized that it is practically
impossible to get SP 5 (i.e. first quality) when dry blast
cleaning a heavily rust pitted steel, but that it is often
possible to obtain when using water abrasive blasting?

In Table 4 of Paper 4 what is meant by 'rust back'? There
are three situations which need to be defined more clearly.

(a) If dry blast cleaning has not removed all the colourless rust producing products they will hydrolyse,
giving black rust spots turning to reddish rust and
corresponding with the rust pits that have not been
properly cleaned.

(b) If the dry blast cleaning has thoroughly cleaned the
surface, condensation will ensue; the surface will
assume an overall fine coating of light red rust
caused by the reaction of water and oxygen (i.e. flash
rust).

(c) If the surface has been cleaned by wet abrasive blasting and all the contamination removed, flash rusting
may occur if steps are not taken quickly to dry the
surface with compressed air or a rust inhibitor is
not added to the water.

Does Mr Keane recognize the two forms of rust back or does
he prefer to paint promptly before the first type of rust
back has time to appear? Does he favour the use of inhibi-
tors when wet blasting and has he studied their effect on
paint performance?

The application specification does not give any warning of
the troubles that can be expected from airless spray application. Have the SSPC encountered such problems as the wide
variation in film thickness which can set up unacceptable
stresses within the coating, and the phenomenon of vacuole
formation which can make nonsense of the expected high film
thicknesses?

Some vinyl primers have a red lead content. The virtues
of red lead are always associated with its specific reaction
and behaviour in drying oil media. Is it really effective
in a vinyl or in an epoxy medium?

Why do the SSPC not list metallic coatings with or without
paint top coats as a method of protecting steel, and why is
there not a category of paint systems suitable for use in
conjunction with cathodic protection?

MR F.D. TIMMINS, Mebon Paints Ltd

Is it possible to formulate a dispersion based primer capable of giving good flow at a dry film thickness of 2.5 mil? Are salt spray tests acceptable in the USA when testing uncoated primers? A recent investigation indicated that primers which gave the best results in salt spray produced the worst failures at cross-hatch areas when tested in complete systems under natural weather conditions.

Rogue peaks produced during blasting may affect durability more than minor residues of chlorides or sulphates. An available alternative to a chemical solution for identifying salt residues on blast surfaces is a thin coat of white PVA emulsion paint which readily indicates salt contaminants by flash rusting.

MR D.F. GOODMAN, British Railways

Table 10 of Paper 4 suggests painting for appearance only when the corrosion rate is less than 5 mil per year. This seems to be on the sparse side. Can Mr Keane comment on the reason for this?

How are supervisors certified in the USA? Who is the certifying authority?

MR M.A. WINNETT, Transport and Road Research Laboratory

I have been working on a reappraisal of the ambient conditions suitable for the maintenance of painting with a view to extending the painting season. In particular, I have been measuring the time of wetness of the steel surface and the temperature of the steel surface. On certain structures, particularly those over dry land, the major factor in prohibiting painting is that the temperature is below 4^0C in the winter for long periods.

Can Mr Keane give details of field trials on painting at below 4^0C and say what systems have been used?

MR W.J. BURLING SMITH, R.J.P. Mickling & Co. Ltd

If sand blasting is coming to the end of its useful life, is the alternative method being considered wet blasting or some other process?

MR I.P. HAIGH, Sir Alexander Gibb & Partners

I assume that specifications of the SSPC are specifications to which one would work contractually and not codes of practice.

CORROSION IN CIVIL ENGINEERING

If American paint manufacturers are willing to supply materials to specifications why are not paint manufacturers in the UK willing, in general, to do the same?

DR J. MORRIS, Scientific Counsellor, South African Embassy

I think the difference between South African standards and the German and American equivalents lies in the method of approach. In South Africa, the South African Bureau of Standards looks after standard specifications of all types. Where a need is expressed, people from public and private sectors and from industry are invited to serve on a committee for setting up a standard. There is therefore a series of paint standards. The nearest equivalent to the German and American standards is SABS 064.[2] This describes the preparation of the metal substrate, the application of paint, the coating system and so on, in terms of specifications for individual paints. In South Africa manufacturers can apply through the Bureau of Standards Marks Scheme for a mark for advertising purposes. They then pay a levy to the Bureau of Standards which has the right at any stage to take samples from their paint production to check against the standard. The Bureau of Standards has its own laboratory, and therefore fulfils a policing as well as a standard setting function.

A committee like that may have a preponderance of manufacturers who may tend to set a standard which suits them rather than the user, but organizations such as the Building Research Institute, which is also represented on that committee, help to maintain the standard of the paint at a level which will ensure that the customer is not cheated. It is therefore possible to set up a code of practice in terms of paints which are specified and therefore controlled.

MR BURLING SMITH

I do not think that corrosion engineers pay enough attention to the design of the component and the structure, particularly with respect to access, from the point of view of application of protection. If certain design features which are recommended cannot be provided, DIN 55 928 calls for additional corrosion protection. How is this to be applied?

In Germany is there a trend to favour aluminium for protection and electric arc spraying?

DR R.R. BISHOP, Transport and Road Research Laboratory

Are temperature limitations and relative humidities specified in DIN 55 928?

MR HAIGH

Mr Hecker states that section 6 of DIN 55 928 '... contains specifications for the execution and control of the corrosion protection work. Among other items, this asks for the quali- fication of the executing personnel ...' Does this mean that the men who apply paint in Germany have a much higher status than those in the UK? In this respect, are they in the same position as welders?

MR B.H. LEIGH-BRAMWELL, W. & J. Leigh & Co.

When a paint composition specification is put out paint manu- facturers must comply with it and give a price accordingly. In this case the paint manufacturer is rather liable to for- mulate his paint down to the specification and put the cheapest materials into the paint to meet it. He can there- fore quote a cheap price. This system allows other manufac- turers new to the market to make paints to specifications without much research work at all. It is possible to make paints to a specification that are completely unsuitable for the application.

Surely a good deal of work should go on in paint manufac- turers' laboratories to get better paint products. If the situation is reached where everything is by specification, there will be no incentive to produce better products and this is not in the long-term interest in the fight against corrosion.

Rather, the user should consider going to a reputable paint manufacturer who has experience in the field and can recommend a system which meets the requirements and has a track record. Such a record must surely be helpful to the person with the corrosion problem.

MR R. ALCOCK, British Railways, R & D Division

British Railways have a surface coatings committee which feeds back practical and maintenance problems to the laboratory, where tests are designed to cater for the situations which normally arise. A coating performance specification with formulation guide-lines is produced and distributed to appro- ved manufacturers and people who ask for it. Manufacturers normally want to sell what they are making, but British Railways purchase enough paint in certain fields to demand what they want.

The first step is to approve the methods of production, quality control of raw materials and batch control of deli- vered coatings of a particular manufacturer. If these are

147

CORROSION IN CIVIL ENGINEERING

satisfactory a sample is accepted to British Railways'
specification. If that is satisfactory the formulation is
approved.

Next a pilot order is placed which takes the form of a
monitored trial on an actual structure. Then one can begin
to see the pattern of the performance of the paint as well
as its application. Then a contract is allocated for the
paint during the period of which the manufacturer must
certify the batch quality control of every delivery. No
change from the agreed formulation is allowed without
British Railways' consent. The engineer can then tell his
staff that a particular paint is applied under certain conditions by a certain method.

This procedure is economic and effective.

MR McKELVIE

DIN 8201 describes types of abrasive blasting. Is there
another standard describing how to use them?

MR KEANE

All SSPC committees on surface preparation, application and
generic types of coating stress open consensus, consideration of dissent, balanced representation, unbiased evaluation
and expertise of committee membership. There are 12 principal
exposure sites at which several thousand test surfaces
representing proposed new SSPC specifications are rigorously
evaluated.

With regard to chemical testing versus appearance, mentioned by Mr McKelvie, it is recognized that neither photographs nor other appearance standards will in themselves be
sufficient for judging surface cleanliness. There are
provisions in the specification for giving some assurance
that the proper procedures are used, and the use of chemical
tests is being considered. British surface preparation specifications are being studied by the subcommittee. In the meantime, I agree that in certain environments and particularly
with certain kinds of coating, chemical impurities can
greatly aggravate the problem of painting.

I think the committee will leave in the brush-off grade.
There seem to be instances where cost is so much of a factor
that one has to choose between brush-off cleaning with a
blast nozzle and hand cleaning or a wire brush, and brush-off
cleaning seems to be quicker, cheaper and better. The choice
is not between brush and commercial, but between brush and a
less thorough method.

I like the potassium ferro-cyanide method and want to look at some others. At present there is a provision concerning rust back. A coating should not be applied when there is any rust visible on the surface, no matter whether that rust originated before or after blast cleaning.

With regard to Mr McKelvie's comments about flash rusting, the development of KUE and other wet blasting methods on a practical basis may trigger this whole operation, because that is one of the real justifications for using it. An inhibitor will lie at the interface between the steel and the primer, which is usually designed to protect without the benefit of such an inhibitor.

I do not know of any paint thickness gauge that will give the answers that are needed. I am dissatisfied with the kinds of surface film thickness gauge that are available. Even the magnetic ones - which may be the most simple and the most practical - have an averaging effect. The report on profile shows graphically that there are about 100 000 peaks in each square inch of surface after blast cleaning. If one has a low film thickness of about ten of them, the surface will show an early failure. Existing film thickness gauges do not pinpoint adequately the very weak points in even one square inch.

With regard to red lead in a vinyl vehicle, recent work indicates that with a vinyl vehicle there is sufficient encapsulation for it not to make much difference what the chemical nature of the inhibitor is. The inhibitor might have some value because it can give good physical characteristics and, if the paint is improperly used or improperly made, one has a safety factor in having an inhibitor when all else fails.

Metallic coatings and cathodic protection are covered in Volume 1 of the Steel Structures Painting Manual, but not in separate SSPC specifications.

In reply to Mr Timmins, the so-called rogue peaks are not rogues really, because they appear so frequently in every small area when viewed under a microscope. Some 3-D pictures have shown that hackles are the main limitation in the kind of media that can be used in blast cleaning. It is hard to get them with sand and non-metallics or with the finer metallics. Under a microscope one can push them down with a probe, but I do not know of any procedure for doing it in production. There is good correlation in the number of these peaks versus early failures, but it has never been proved that the failure was at that particular locus. However, I think there is a good presumption of cause and effect.

As to the salt fog method, a better method than the

accelerated methods which are available is being sought -
one that will show in a reasonable length of time what the
coating will do.

Thinner water based films have given a greater succession
of early failures, and it is found empirically that at
2-2.5 mil the early failures are reduced. That is one of the
many criteria needed in a water based paint.

A dry film thickness of 2.5 mil is obtained by skilled
formulation and skilled application.

With reference to Mr Goodman's remarks, I was trying to
illustrate the concept that there are times when one
would paint for appearance only, and that there is a dividing
line. I think that 5 mil is really a little on the high side,
but it is a criterion that has been used. I do not hold with
all the details of that criterion, but I do hold with the
broad concept that one should look at the real necessity for
painting.

The nuclear power industry is the only one I know of that
is just beginning to have certification of inspectors and
consultants. There are other certifying bodies in several
States, including California and Florida, where one can be
listed as a corrosion engineer or a corrosion specialist, and
the National Association of Corrosion Engineers has a similar
set of categories.

Mr Winnett asked for details of field trials on painting
at below 4°C. The systems tried in this particular series
were rather limited. For alkyds, oil based, and combinations
thereof, results indicated considerable tolerance for poor
application conditions. There were interactions of temperature, wetness and surface cleanliness, and unless all three
factors were working against one, better results were
obtained than one might expect. The results of that work
are not overemphasized because this tolerance tends to give
the applicator a false sense of security. The report will
show that one has a little more tolerance than used to be
generally supposed. SSPC ordinarily specify that the surface
must be dry and above the dew point in temperature.

In reply to Mr Burling Smith, in addition to wet blasting,
other media, including abrasives low in free silica, and
machinery which will allow the use of metallic abrasives are
being considered. The wet methods, including the KUE water/
sand/air combination, look promising. Water blasting and
flagellating types of wheel are also being considered. The
scanning electron microscope shows why this kind of surface
does not seem to provide the same substrate for long-term
protection. Attention is also being given to a few chemical
methods and disposable abrasive media which can be used in

such places as in the holds of ships.

With regard to Mr Haigh's remarks, there are important semantic differences between the various terms such as code of practice and specification. In the latest review the term specification is retained because it is felt that this more nearly describes the provision for the procurement of materials for a particular paint or paint system. The American Society for Testing and Materials, for example, avoids the term specification, in favour of standards and operating practices (which are like codes of practice in the UK). A few hundred thousand copies of our surface preparation specifications are used every year, many of which are embodied in the contract or procurement document. (The SSPC materials specifications are often used in the same way.) They are the official surface preparation specifications of the American National Standards Institute and have been referred to as 'the standard' by such organizations as the American Association of State Highway Transportation Officials and the Canadian Government Specifications Board.

On the subject of the consensus including the main paint manufacturers, this was a long time coming and it probably has not come all the way even yet. Committees have an open membership, and as a result there are representatives of the outstanding paint manufacturers on each of the 20 principal committees. Here they battle it out with the ultimate users, with raw material suppliers, with applicators and with representatives of the public interest sector to make sure every specification is fair to all. As a result some of these specifications require many years of development.

The same trends are seen towards proliferation of specifications in the USA. Each State has its own highway department specifications. In each State there are specifications for counties, cities and other subdivisions. However, there is a trend towards consolidation in specifications, e.g. the Federal Government has abandoned some specifications to let - private consensus standards prevail wherever they can.

Consensus standards enable procurement on an open competitive basis. They are not intended to squeeze out the proprietaries, many of which will conform with the specifications, which now include as many performance provisions as possible. However, neither composition nor performance requirements are always enough to guarantee that the product will do what it is intended to do, because there is always a temptation to formulate to low cost and to just pass the specification. Not so much in the paint specification but in the performance-oriented paint system, the concept of the track record or case history of successful application and

CORROSION IN CIVIL ENGINEERING

use has therefore been introduced.

MR HECKER

In reply to Mr Burling Smith, when certain design features
which are recommended cannot be provided, either additional
coats are applied before assembly or a protection system with
higher performance is applied. Aluminium spraying for surface
protection is used only for some special purposes. It has no
importance for ordinary protection.

Dr Bishop asked about temperature limitations and relative
humidities. In part 4 of the DIN standard it is expressed
that the object temperature has to be without doubt above the
dew point. In part 6 methods showing how the dew point is to
be determined are given. Minimum temperatures for certain
coating systems are explained in part 5 of the standard
(i.e. 10°C minimum temperature for two-pack epoxy systems
and 0°C for two-pack polyurethane).

In answer to Mr Haigh, for certain corrosion protection
application, paint contractors have to apply for official
approval and prove that applicators have been trained
specially for the task (i.e. interior lining of tanks and fire
retardant products and application in nuclear power plants).
In these fields there is a special training with certificate.

In reply to Mr McKelvie, DIN 8201 parts 1-10 describes the
different types of abrasive. There are no other standards
which describe how to use them. Other references to blast-
cleaning of steel surfaces are made in Swedish Standard SIS
05 5900 and British Standard BS 4232.

In reply to Mr Leigh-Bramwell, if one has 50 customers with
50 different specifications, then it is worthwhile trying
to get them concentrated on one, which is DIN 55 928. I would
not go as far as to produce a formulation, but within certain
limits regarding the minimum content of certain raw materials,
if the paint is named after the raw material I think it should
be a good guide, and should provide at least the same opportunity of competing with other paint manufacturers. I think
it is worthwhile instituting certain regulations for that.
However, it is costly to manufacture different products
according to the specifications of different customers for
the same purpose and with the same performance.

References

1. Swedish Standard SIS 05 5900, 1967.

2. South African Bureau of Standards. Preparation of steel
   surfaces for painting. SABS, CP 064.

Discussion on Papers 6, 7 and 8

MR BARTLETT

On the contractual side, never forget that a contract is an agreement between two parties to do something to their mutual advantage. It is a relatively simple business, one might think, if one is buying a bus ticket, although legal case histories indicate otherwise. It is more complicated if one is buying a bridge or a tunnel, and the only way that one can get a true meeting of minds on such a project as that is if there is a well established way of achieving the objective. Paper 6 shows that some corrosion protection details are often best left to be arranged later.

MR MOBSBY

At present it is the exception for an applicator dealing with general structural work to have adequate quality control facilities. Hence, if a client wants a good job, he is left with the necessity of providing his own full-time inspection. In doing so, he may assume responsibility for the protective coating and, because of recent legislation, become enmeshed in product liability. To overcome this, I suggest more emphasis should be placed on quality assessment of applicators before work is let. Unless applicators have proper facilities, discussion of paint system details is wasted.

MR SMITH

The economic question that engineers dealing with corrosion ought to be always asking is, 'To what extent does it pay to spend money when a structure is built, in order to minimize maintenance later?' When interest rates are high, it might often be difficult to justify any cost for initial protection. The day may well come when economists will be saying to government that it would be more economical to build cheaper, and to rebuild more often. Engineers need to be able to argue this point on equal terms with economists and to challenge their basic assumptions. If economists are wrong, engineers must tell them so, and do it convincingly.

MR E.M. GOSSCHALK, Sir William Halcrow & Partners

I believe that performance specifications can provide a sensible approach with only a short guarantee period, perhaps of just one year, in service. Inadequate application and formulation often need less than one year to make themselves evident, and long guarantee periods are of dubious value. Nevertheless, performance specifications may be ruled out when prospective owners want to use protective systems which have already served them well. Owners may wish to standardize just to avoid having different systems to maintain. I believe it is essential to include clauses in a specification which report the service conditions which the protected installation is expected to endure, which require that the protective system shall be applied and cured in accordance with the manufacturer's recommendations, and which call for evidence of satisfactory use of the protective system offered under similar service conditions.

Consultants and prospective owners need to look particularly carefully at the evidence that new formulations are sufficiently well proven. A slight change in formulation can transform a satisfactory paint into a completely unsatisfactory one without a change in trade description. It is also important not to adopt a solution which may have worked well elsewhere but not under the same service conditions.

Figures 1 and 2 show the casing and flange of a submersible well pump to a reputable manufacturer's standard design and protective treatment which within twelve months reached a state of complete failure. Holding down bolts corroded completely and bearings became displaced.

Figures 3 and 4 show a galvanized rising main which similarly failed due to pitting, corrosion being particularly evident along the line of a cable which ran up the pipe. In another well field in the same region, where conditions were superficially similar, no such problems occurred.

The trouble has been variously ascribed to air entrainment in the water, lack of circulation of heated water adjacent to the pumps, use of dissimilar metals and to microbe population in the water. The lessons to be learned are that nothing must be taken for granted; every consultant must be a

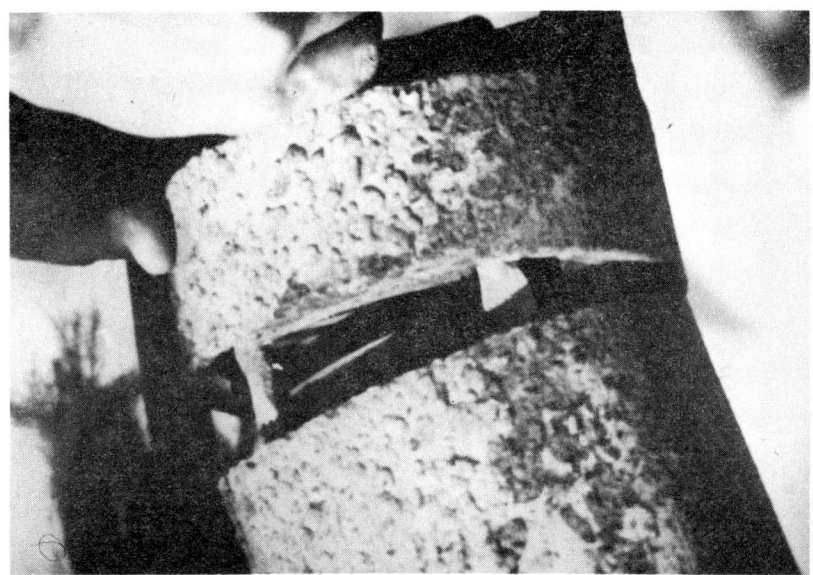

Fig. 1

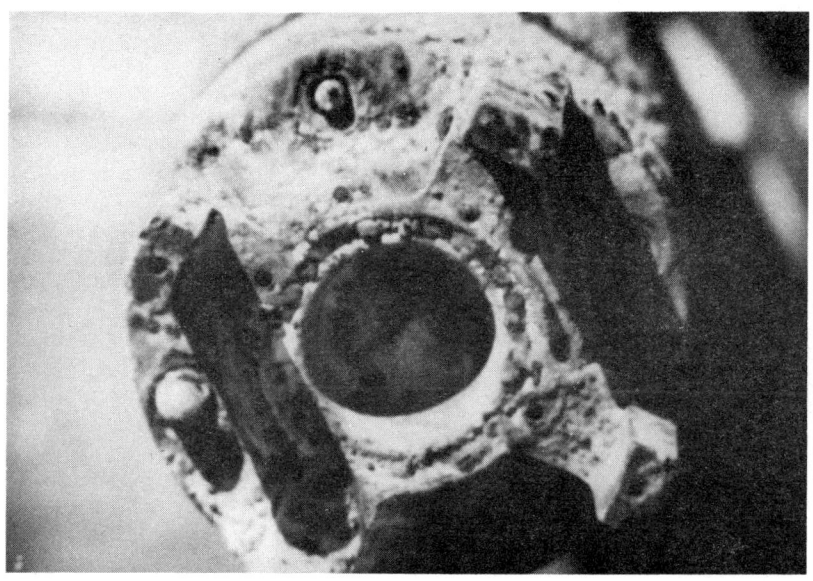

Fig. 2

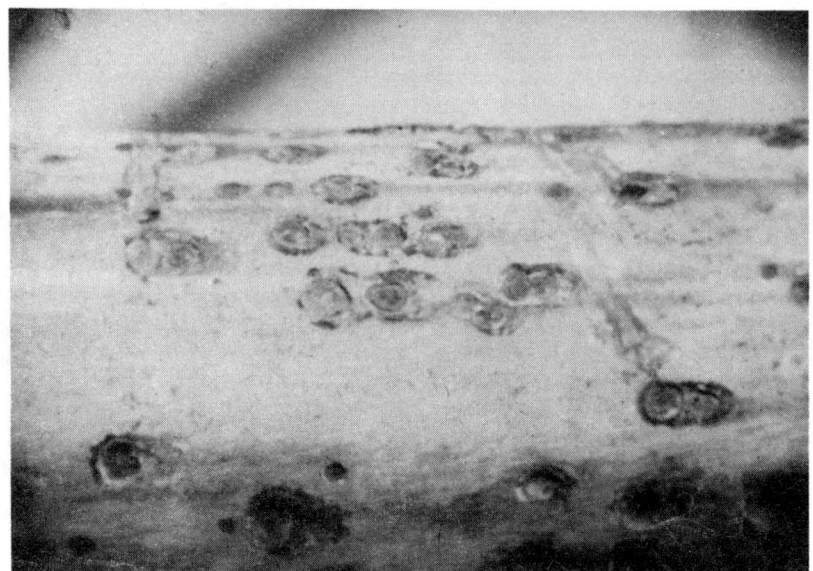

Fig. 3

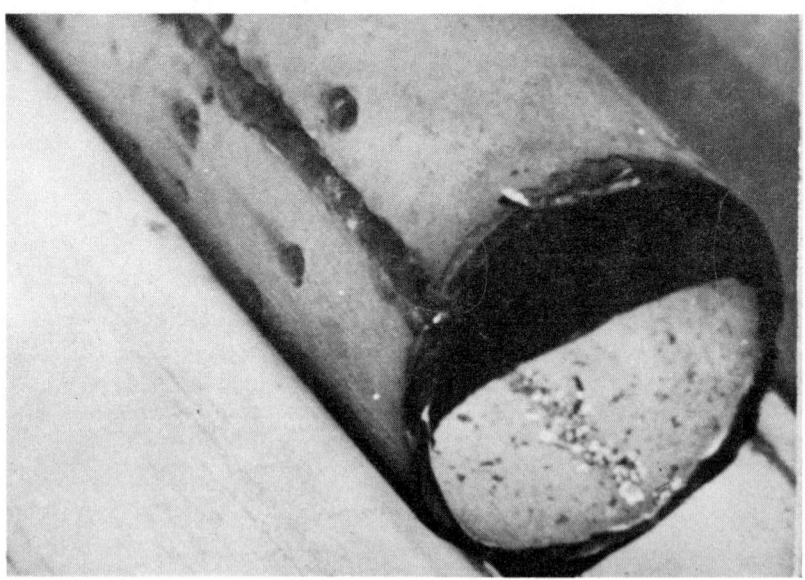

Fig. 4

corrosion consultant as it is impracticable to engage specialist firms at every turn.

For overseas projects, an important consideration is suitable protection during shipment, when conditions may well be much more severe than in final service. I know of a case where a zinc spray treatment failed on the voyage from Liverpool to West Africa. BS 5493 would give such a coating of life of, say, five years in contact with salt spray and sea water. One is warned that the recommendations of BS 5493 are related primarily to conditions in the UK, but this experience suggests that one could be seriously misled.

Contamination by other cargo is also a hazard. It may be necessary to specify shipment below deck and prescribe a suitable protective covering that will not become saturated or contaminated.

Generally, coatings to be applied at remote sites should be as insensitive as possible to inexpert treatment and to unkind atmospheric conditions. In particular, treatments which need warm, dry curing conditions should not be used in cold, damp places such as tunnels and pipelines.

With regard to economics, I believe the need is for commonsense and to substitute engineers for economists.

MR J.B. BODEN, Transport and Road Research Laboratory

I would like to show some of the implications of short-term solutions and the wisdom of providing for the long-term future. In 1979 the replacement value of the existing 130 000 highway bridges in the UK is about £8000 million, the public sewers probably approach £30 000 million, the water mains approximately £15 000 and the gas mains say £7000 million: a total of £60 000 million for just four items. Supposing these services are all well constructed and, on average, they last 100 years. Then just to replace the decay, excluding maintenance, could cost £600 million a year. I prefer this simplistic way of looking at the implications of short-term solutions to discounted cash flow.

MR D.F. GOODMAN, British Railways

I think it is important to realize that any discounted cash flow calculations or net present value calculations depend on information provided by engineers and corrosion technologists. If therefore the calculations give silly answers, the reflection is on them and not the economists.

CORROSION IN CIVIL ENGINEERING

MR A.N. McKELVIE, Paint Research Association

British Standards documents deal with quality assurance but
mainly for fabrication; they do not mention protection. I
believe the protective function must be integrated with the
fabrication function to give a good job. Large contractors
need two protection experts in their organization: one in the
design and planning department, looking after protection
functions from the start, from the cardinal date plan down to
the detailed planning, and the other in the production
department, in charge of all the functions relating to the
protection of the structure. The latter must be responsible
for providing the proper equipment for application and
environmental equipment at the right time either for his own
work force or for any subcontractor. If all this is done, I
believe there is less need for thorough inspection, although
the ultimate responsibility must rest with the main contractor.

MR I.P. GILLSON, Central Electricity Generating Board

In the CEGB, quality assurance is oriented mainly towards
fabricators. One of the main provisos of the system is that
quality inspection must be independent of production management and reported to senior management. A firm that has
passed a quality audit and received a three-year certificate
has the facilities available to produce the good work. It
does not necessarily mean that he will apply it, so the audit
is not a substitute for inspection. However, it is a useful
prerequisite for the production of good work and lessens the
need for inspection.

DR J. WEAVER, Sir William Halcrow & Partners

Does the Authors' concentration on the impressed current
system indicate that they prefer it to the sacrificial anode
method, and if so, why?

    As I understand it, the sacrificial anode method is suit-
able if an electricity supply is not available, if electricity power costs are high or if the resistance of the
electrolyte is low, as in sea water. The cost of maintaining
the system is low, but there is no convenient method of monitoring its performance. The impressed current system
requires a continuous electricity supply, is suitable for
highly aggressive conditions and where the resistance of the
electrolyte is high. It requires maintenance and inspection,
but its performance can be monitored.

    Could the Authors give a fuller comparison of the two

Fig. 5

methods, including a comparison of costs, and an amplification of their advice to consider the alternatives of protective coatings and an additional metal allowance? The use of cathodic protection in modern structures is extensive. Can the Authors give details of such applications, in particular the efficiency of current distribution to structures, from the anodes of impressed current systems? How can one avoid the spark or arc hazard which can result when the electrical bond between vessels and some cathodically protected structures is broken?

I have considered whether concrete containing some moisture could be deemed an aqueous solution, in the sense used by the Authors, and whether particularly important steel elements encased in concrete, such as prestressing tendons, could be protected by cathodic protection, if required.

Figure 5 shows steel beams in a marine structure of composite construction of steel and concrete which is about 25 years old. The bottom flanges of the beams have been subject to severe corrosion, and the concrete of the beams has a high salt content from splash and spray. I deduced from examination of the beams that the corrosion was electrochemical and wondered whether it could have been prevented by cathodic protection. The concrete was cut away using high pressure water jettings, and the corrosion products were removed by the water jet and grit blasting. The electrolyte – the concrete – could be provided with a controlled porosity by air entrainment, but the problems of its discontinuity where cracks occur and where protection is most required, seem to present significant obstacles. The secondary reinforcement and problems with bonding between elements also seem to present obstacles. What are the Authors' views on this?

MR I. HROMATKO, Institut Geoexpert, Zagreb

My company is involved in the investigation, design and execution of, among other things, cathodic protection systems. Cathodic protection is a useful method for anti-corrosive protection of steel structures, underground and in water. My firm has used cathodic protection to protect the external and internal surface of steel water wells and submersible pumps in them, heat exchangers, steel piles, steel in reinforced concrete structures underground and in water and steel anchors. Civil engineers are beginning to count on cathodic protection as one of the methods for anti-corrosive protection; it is one method that can be used after construction and all elements of it can be regulated after its application.

For well-coated steel pipelines, less than one million per square metre are needed and for poorly protected or painted steel pipelines, about ten million per square metre are needed. For bare steel structures about 100 million are needed and in a special case in a seashore area in Yugoslavia 150 million per square metre are needed. It is very reasonable to give primary protection of underground or underwater steel structures and this can be done with any kind of paint or coating, bituminous or organic material.

DR R.R. BISHOP, Transport and Road Research Laboratory

Full-scale experimental cathodic protection installations of bridge decks have been reported by Stratfull[1] and Fromm[2].

MR I.P. HAIGH, Sir Alexander Gibb & Partners

Corrosion of reinforced concrete bridge decks is a much more serious problem in the USA than in the UK. For many years it has been the practice in the UK to waterproof bridge decks. Reports of attempts in the USA to drive chlorides out of concrete bridges electrochemically indicate a policy of desperation.

MR SMITH

Oversimplifying slightly, steel piles driven into undisturbed ground below the water table do not corrode, and do not need protection - cathodic or otherwise. Above the water table one could not get the current needed for cathodic protection anyway. However many sulphate-reducing bacteria there are, one can generally drive a steel pile into undisturbed ground and it will be safe.

MR R. ALCOCK, British Railways, R & D Division

What are the merits of steel piles unprotected, steel piles coated, and steel piles cathodically protected, against concrete piles in a tidal situation where there is a neutral pH value but a high chloride content?

MR M.M. MILLER, Sir William Halcrow & Partners

What quality of paint treatment is required in combination with cathodic protection to reduce the running costs due to the current required? In desert conditions are there any problems with solar panels due to obscuration from windblown sand? How often is it necessary to top up batteries in these remote places?

MR M.W. CRABB, Sir William Halcrow & Partners

What are the problems involved with wrapping tapes used to
protect the pipes? Did the Authors encounter any problems
of disbonding?

MR M.J. McCULLOUGH, Wimpey Laboratories

It seems generally accepted by engineers that, when using
Wenner spreads, there is a 1:1 relationship between depth of
penetration and electrode spacing. This assumption can give
erroneous results. In fact the relationship is far more complex.

Also, resistivities from just two electrode spacings
(e.g. 2 m and 4 m) are often quoted as the soil resistivities.
This is not true; they are the apparent resistivities of the
soil. To obtain soil resistivities it may be better to do an
expanding Wenner spread and interpret the field curve. This
will give a better representation of the soil resistivity and
depth sounding.

MR W.J. BURLING SMITH, R.J.P. Mickling & Co. Ltd

With increasing attention being given to handling hot products and with compressor stations, do the Authors consider
that the negative 850 potential is still adequate protection
when the steel is heated?

MR T.V. PERROT, Department of Transport

One has to be careful to confine the use of any steel -
weathering steel in particular - to its particular and acceptable environmental limits. For highway structures it is
recommended that unprotected weathering steel should not be
used in marine locations or when it may be subjected to de-
icing salt spray; this is because of the possibility of the
development of unacceptable corrosion pits.

The Authors warn against the lack of self-protection if
the steel is used in a continually wet situation. I suggest
the designer should ensure that all parts of the structure
are able to dry out and that the possibility of joints
leaking should be allowed for in the design and positive
drainage provided.

Could the Authors comment on the sustained decline in the
rate of corrosion with respect to the exposed and sheltered
situations? Do they consider any long-term programme, say
over 60 years, is required to monitor and confirm the

predictions?

Further to paragraph 17 of Paper 6, for Department of Transport structures contact should also be made with a technical approval authority.

The appropriate added thickness to allow for corrosion loss is indeed difficult to decide. Projected values for the severe exposure at Tinsley Viaduct, based on the 120 year life of the structure, would give about 4 mm per surface for the open exposure and a lower figure for the sheltered. However, it has been suggested that these figures are excessive, assuming there is some stabilization of the surface. Consideration may be given to a 2 mm allowance per surface, backed up by a thorough maintenance inspection system; this would reveal a 1 mm allowance per surface in a mild environment.

About 40 highway structures have been constructed in weathering steel, the largest of which used 1800 tons of weathering steel. Investment of this magnitude confirms that there is confidence in this type of steel.

MR GOODMAN

On the River Iden Bridge, there is a marked contrast between the appearance of the weathering steel on either side of the trimmer girder. Why should this be so?

Can the Authors give any corrosion rates for weathering steel in a difficult area such as that on the abutment side of the trimmer girder?

MR R.F. MANDER, Department of Transport

I think it is inevitable that staining or weathering will be uneven; at best this will lead to the motley appearance of a bridge, but it could cause significant loss of metal around the webs of the girders near the bearings - an area that is constantly subjected to de-icing salts in solution. How is the structure to be painted when this stage has been reached? Should one just paint around the bearings with a matching paint or a contrasting paint, or should one grit blast the whole structure down to white metal and then paint?

During construction, cement and grout are bound to drop on to the beams. How should one clean this off and should one clean the whole of the structure to get a nice even colour?

DR J. MORRIS, Scientific Counsellor, South African Embassy

My experience has been that horizontal flanges and sections

rust more quickly than vertical ones. In determining the
rate of rusting, has this been taken into consideration?
How can one compensate for it in design?

In South Africa manufacturers of roofing sheets made from
Cor-Ten steel have relied on the greater strength of Cor-Ten
and reduced the thickness in order to bring it into the
correct price class. This seems to me to be a contradiction
of what the Authors of Paper 8 recommend.

MR G.K. WOOD, Messrs Sandberg

The cost of steel maintenance is staggering, so the money
available must be used to give the best possible protective
treatment.

To achieve quick results I think it is vital to give more
attention to plant and people.

Plant is apt to be very old, overloaded and subject to
breakdown. It is old because the last part of the works to
be modernized is generally the protective treatment unit.
It is overloaded because the works were often originally
designed for less throughput. Thus when a job is running
late there is a panic to get it through somehow. This tends
to make for a breakdown of plant and further delay.

Thus for the first priority I would ask for plant improvement as quickly as possible.

As regards people, I think blasting and painting are
probably jobs that are only taken up as a last resort. Blasting is a tough unpleasant job and I think blasters could be
given encouragement if upper management would go and see for
themselves where improvements could best be made. The men
would see they were not a forgotten army.

This is a lead to quality assurance which is a good
answer, but I think it will take a long campaign to solve the
problems.

MR E.N. CAPPITT, Posford, Pavry & Partners

Consideration must also be given at times to existing structures having inaccessible corroded members. After the glass
and timber were stripped off the main frame of the Temperate
House at Kew Gardens, which the Department of the Environment
wished to preserve, considerable corrosion and corrosion
products were found inside the structural cast iron rain-
water pipe columns. Is there any satisfactory and economic
method of cleaning and protecting such members for which
wire brushing and sand or water blasting are not appropriate?

MR E.G. ROBSON, Advanced Sealants Ltd

Some chemical compounds which are advertised particularly for motorists can be applied straight on to rust. It is claimed that they react chemically with the rust, form an inert substance, stop further corrosion and form a primer ready for painting. Such methods seem very attractive because they avoid expensive and tedious cleaning and preparation of corroded surfaces. These compounds are based on phosphoric or other acids and some have zinc in them.

Have such chemicals been used on larger structures? Do they do what they claim?

DR BISHOP

Rusty steel formed in a marine or industrial environment cannot be effectively treated with rust converters. However, rusty steel formed in a rural environment can usually be effectively treated by wire brushing and subsequent paint treatment.

MR BARTLETT

There are great advantages for many structures in spending money on initial treatment. One can do it properly then. It may be extremely difficult even to get access to do it later. It is a big factor in the economics and money which has been specially authorized as capital cost should be used in this way; money for maintenance may be much harder to obtain. The engineer should seek to be in a position to ensure sensible decisions in this matter.

MR J.E. O'BRIEN, British Steel Corporation

British Steel has carried out extensive work on the investigation of corrosion in steel sheet piles and steel bearing piles. Generally each case must be investigated on its own merits. In general

  (a) corrosion below mud level on driven piles is negligible

  (b) below low water line corrosion is very slow, but the presence of pollutants may cause problems at the low water line

  (c) splash zone corrosion is the worst and this is dependent on site conditions such as weather, wind and tide range.

CORROSION IN CIVIL ENGINEERING

Corrosion can be tackled in a number of ways

(a) designing for mild steel and constructing in high yield steel
(b) increasing the thickness of piles where possible
(c) adding copper to steel to reduce splash zone corrosion by up to 50%
(d) painting with coar tar epoxies or tars
(e) using low alloy steels resistant to corrosion; these are actively being investigated.

MR L. WOOLF, Metrotect Ltd

My company manufactures coal tar and bitumen for pipeline protection. Coal tar, and to a lesser extent bitumen, have been used on many thousands of kilometres of pipeline throughout the world. There are about 2000 miles of British Gas line and over 2500 miles in the North Sea. The land lines are protected by impressed current and the submarine lines by sacrificial anode cathodic protection. Overseas many thousands of lines have hot applied coatings on them and to my knowledge there has been no breakdown. However, there have been ten year failures in the Middle and Far East on cold applied tapes, which have degraded with serious results and high cost to the operator. My company is currently doing research on cathodic disbonding.

Paper 7 refers to an epoxy powder coated line. How much damage was there to the coating when the pipe arrived on site, and how was this repaired?

Are the Authors happy that there is no cathodic shielding in these areas?

For pipe to soil potential, how do the epoxy powder, the bitumen coated lines and the 30 km of coal tar coated lines compare? On the bitumen coated water pipeline was there any root growth through the bitumen coating on to the pipeline surface?

MR R. KING, Charles Haswell and Partners

My comments relate to work in the Middle East.

My firm has Arab clients who demand engineering perfection but are not particularly receptive to engineering niceties. In the Middle East corrosion rates are far greater than in the UK. For example, pitting corrosion rates in mild steel pipelines are measurable in millimetres per month rather than microns per year. To reduce maintenance, therefore, there is a tendency to substitute more expensive engineering materials, e.g. the non-ferrous metals and stainless steel.

On multi-disciplinary plant projects it is important to
interrelate the design plant life with that of the civil
engineering elements. It is therefore important to be well
informed on the lives of thin film paint systems. More information on the behaviour of such systems in the Middle East
would be very helpful.

It would be interesting to know more about the behaviour
of concrete as a protective medium to prevent corrosion of
mild steel structures. My firm is currently considering concrete lining to the inside of mild steel pipes as an alternative to thin film barrier coatings, the latter not having
stood up well to the aggressive service conditions.

MR R.J. SERVICE, Property Services Agency

In Ministry of Defence naval bases where vessels come alongside for refits surplus rubbish often gets thrown overboard.
Where this debris exists what guarantee is there that an
impressed current system monitored above water is protecting
the structure as opposed to the metallic debris?

MR BARTLETT, MR SMITH AND MR MOBSBY

Mr McKelvie and Mr Gillson make the point that good production facilities (which include planning and quality control)
lessen the need for inspection. This is entirely true
However, few general fabricators possess such facilities,
and they are unlikely to do so until they receive sufficient
work to warrant their provision. Hence, at present the
client must provide independent inspection.

Mr Boden gave useful figures on the enormous capital cost
of the infrastructure. At present maintenance of bridges in
Scotland is approximately 12½% of what it will have to
become, on the assumption of a 120 year life, if the stock of
bridges is not to fall into decay. This is not just a money
matter: it is a manpower problem, demanding a change of
attitude. Structural engineers are going to be increasingly
concerned with maintenance, and less with construction. This
is a point which economists cannot get over.

However, the economists cannot be bypassed as government
is going to listen to them. Engineers must convince economists; it is not sufficient for them to convince one another.

We agree with Mr Wood. A good maintenance man is a
different kind of person.

MR ALLEN AND MR LEWIS

In reply to Dr Weaver, we gave the impression in our Paper
that we preferred power impressed systems because the only

three systems we quoted were of that type. The choice
between the two systems is on the basis of cost and convenience. Interaction in an urban area may be considerably
reduced by using the sacrificial anode system. Convenience
of maintenance is probably the most overriding consideration.
Power impressed cathodic protection systems need much more
frequent monitoring and maintenance and this is often
unacceptable. Certainly cathodic protection is extensively
used in harbour structures, offshore structures, foundation
piling and large diameter pipework (internally). Current
distribution depends on the nature of the coating on the
surface to be protected and the resistivity of the electrolyte.
In sea water, there is no problem in spreading protection
extensively on to a well coated structure.

Concrete may be regarded as an electrolyte. However, one
should look to methods other than cathodic protection for
protecting reinforcing concrete. Adequate cover to the
primary and secondary reinforcing is the first necessity.
Aggregate size in relation to cover must be controlled. Only
chloride-free ingredients should be used and by mix, design,
careful pouring, compaction and curing, the best possible
concrete should be produced. In these circumstances, reinforcing corrosion problems would probably disappear. Although
concrete may be an effective electrolyte, current distribution
becomes a problem because there is a narrow cross-section for
conduction on to the reinforcement. The result is that
cathodic protection would be an extremely costly method of
prevention.

On the choice between sacrificial and power impressed
systems in the North Sea, for example, where extremely large
structures are often uncoated, there is a preference for
impressed current which is readily adjustable. The problem
is that systems must be extremely robust and this is more
easily achieved, in the present experience, with sacrificial
schemes. Another advantage of sacrificial schemes in the
North Sea is that the structure becomes protected immediately
it goes into the water.

With regard to arcing and sparking, certainly there is a
hazard with cathodic protection currents in inflammable
atmospheres. The methods usually employed to overcome static
electricity problems are appropriate and procedures are
mentioned in, for example, the Institute of Petroleum's code
of safe practice.[3]

The American problems seem to arise from salt deliberately
introduced on to highways for de-icing rather than from sea
water or atmospheric salt being absorbed into the concrete.
Cathodic protection is the last method of prevention we

would use in this situation.

In reply to Mr Miller two figures were mentioned in our Paper to illustrate the difference between current demand for protection of bare steel and coated steel. Using the costs of coating and of power, it is possible to produce an optimum design. Solar units in desert locations must, of course, be sited where they are not going to be overtaken by advancing dunes and must be protected from sand accumulation. The frequency with which batteries must be topped up depends on the degree of control of the voltage applied to the batteries, but it is roughly about once a month. Manufacturers do not appear to be concerned about obscuration of the panels by sand. The small amount of dust that settles on the panel does not seem to affect the output of the panel.

With regard to disbonding there are problems with tapes and with all other coatings subject to cathodic protection. There are tests which are used extensively by the British Gas Corporation to determine the disbonding propensity of coating materials. Our experience has shown that it is necessary to reduce the maximum negative potential employed on tape coated pipelines compared with coal tar coated pipelines. The effect of this is that one may obtain an improved current spread if a greater value of negative potential is allowable at the points of application.

We did not say that we prefer wrapping tapes to other types of protection.

In reply to Mr McCullouch, ordinary Wenner traverse as is normally used for this sort of work certainly does not give the actual soil resistivity - it is a convenient method of assessing route corrosivity. It is not really necessary to know the actual soil resistivity.

If the NERNST formula is used for deriving protective potential, it appears as though the value on heated fuel oil pipes, for example, may increase marginally from 0.85 to about 0.87. We have not seen yet any original work published on this. As the general specification for performance of a cathodic protection system may require a margin of safety and sets a limit at 1 volt anyway, this problem has not been seriously considered.

In our experience, one cannot generalize about the merits of protection because different situations appear to have different potential corrosion rates. There have been cases where a decision on the application of full cathodic protection has been delayed for two years pending the outcome of trials. Piles were drawn after three years and examined for corrosion. In one case it was decided to proceed with cathodic protection but in other cases it was considered that there was no need. The critical area in piling is very often above water level which is the area which requires corrosion prevention methods based on materials' selection rather than cathodic protection.

Little significant damage was found to the powder epoxy coating mentioned in our Paper. Pipes were shipped and transported with protective rubber straps at two or three positions so they were separated from impact. Minor damage was repaired by the sealing wax method of dabbing on powder and sintering it with a flame. However, the welded joints were coated using a mastic tape.

The coating showed an average current demand of 50 $\mu A/m^2$ which compared favourably with coal tar enamel coated pipelines. This is a recent project and there is no information available on long-term current demand. We restrict the maximum negative potential to 1.5 V which is the level taken anyway for tapes. Attenuation was certainly as good as coal tar. Experience may well show that higher voltages may be used on powder epoxy without attracting disbondment problems.

In reply to Mr King, it is not usually economic to use cathodic protection in small diameter pipes. In general, it is also more expensive in large diameter pipes than concrete linings.

DR KILCULLEN AND MR McKENZIE

We would agree with Mr Perrot that there are environmental limits on all materials and these should be observed. Salt spray and continually wet situations are certainly to be avoided when using Cor-Ten steel.

On the subject of the reduction in the rate of corrosion of the steel, this is closely linked with the conditions of exposure - particularly with the degree of sheltering. It would therefore be sensible to monitor Cor-Ten performance on actual structures until a clear picture can be built up. In fact, work is in hand to do this using a recently developed ultrasonic technique which can measure the residual steel thickness without any disturbance of or interference from the rust film. We would accept, in general, the points Mr Perrot makes about the corrosion allowances for the Tinsley Viaduct site, but would add one cautionary note: open exposure is known to give higher initial rates followed by a greater degree of reduction than sheltered exposure and so any predictions of the 120 year performance based on these are likely to be misleading.

The relationship between appearance and corrosion rate,

raised by Mr Goodman, is complex. Certainly at Iden there
are quite big differences in the appearances of the rust
at various points on the structure. However, we think it
unlikely that the appearance of the rust always bears much
of a relation to the corrosion rate. We have certainly found
that corrosion rates and appearance vary a lot, but there
does not seem to be any clear correlation between the
appearance and corrosion rate. The difference in the appearance is probably because there are, in effect, lots of
different climates on the bridge. However, we would strongly
advise that, where there is a marked difference in appearance
between two areas close together, the situation should be
investigated. The new ultrasonic device would be ideal for
this. It seems likely that this particular problem at Iden
Bridge has resulted from leaking expansion joints. There is
possibly a supplementary point here on the question of
inaccessible areas. It does not really matter whether or not
one paints inaccessible areas initially because if areas
cannot be reached for maintenance there is going to be a
problem. If there is no way of designing out an inaccessible
area, thicker steel must be used to cope with corrosion.

There are no rigid answers to the problems raised by
Mr Mander. If the problem is obviously a local one due to
some design feature, then painting of that area would be a
possible way out. In fact, BSC have work in hand on the
choice of paint schemes for application over weathered Cor-
Ten. Tests are being started on Cor-Ten samples that have
rusted for ten years, and the possibility of applying paint
schemes directly over the rust after wire-brushing will be
investigated. There is some evidence that because of the
different nature of the rust, a paint scheme could be successfully applied over it.

The answer to the problem of staining of Cor-Ten steel with
other building materials depends on the length of time one is
prepared to wait for the situation to be put right. If the
stain is blasted off then uniform rusting will be obtained
relatively quickly. However, if one is prepared to wait
several years the stain can be expected to weather off
naturally. Relatively large deposits of building materials
on Cor-Ten steel should be washed off if they are still wet or
chipped off later.

We agree with Dr Morris's views on the rates of corrosion
of horizontal and vertical pieces of Cor-Ten when these are
freely exposed in the atmosphere. However, what little evidence we have, mainly from Iden Bridge, suggests that the
difference is not so pronounced on the undersides of bridges.
On the question of the roofing sheets, all we can really say

is that we would not recommend using the practice Dr Morris
describes in the UK. The situation may be very different in
South Africa because of the corrosion rates and atmospheric
conditions there and performance may be more akin to
performance in the USA. Certainly the sort of area we would
be most wary of in a situation like that would be the joints
where any moisture that does get on to the surface tends to
be trapped.

References

1   Stratfull, R.F. Experimental cathodic protection of a
    bridge deck. Highw. Res. Rep., 1974, Jan., 19501-76503-
    635117.

2. Fromm, H.F. Cathodic protection of re-bars in concrete
    bridge decks. Material Protection, 1977, vol. 11, 21-29.

3. Institute of Petroleum. Electrical safety code, 5th edn.
    Applied Science Publishers, Barking, 1965.

Discussion on Papers 9 and 10

MR F.D. TIMMINS, Mebon Paints Ltd

Paper 9 concentrates on hot dip galvanized steel. During my
48 years at British Rail, I acknowledged the contribution of
galvanized steel in protecting railway carriage roofs, merry-
go-round wagons and electrification overhead structures, the
traditional problems of paint adhesion being reasonably
reduced by using a mordant T wash developed by British Rail.

Unprotected galvanized steel exhibited extensive failure
when subjected to continual acidic fall-out from industrial
areas. Lightweight overhead gantry areas with thinner galvanized coatings were severely attacked, but even the heavy
duty masts became corroded in due course. In sharp contrast
the sheltered parts of the structure developed a basic zinc
salt patina.

Extra paint protection for all galvanized areas likely to
be inaccessible or subjected to industrial prevailing wind
fall-out is advisable.

Extensive practical experience with metal sprayed coatings
has varied considerably from the technical sales literature.
Arc sprayed aluminium invariably produces a spongy texture
and is difficult to apply uniformly because the operator's
vision is obscured by dense white fumes.

Gas sprayed aluminium and zinc rely on critically prepared surfaces and, being sensitive to alkalis or acids,
rely heavily on chemical resistant paint protection.

Notable failures have occurred with metal sprayed coatings. British Rail's Hook Bridge protected by aluminium
failed due to saponification of the conventional micaceous
iron ore paint by alkaline road seepage and likewise the
Department of the Environment's Boston Mann Bridge failed
due to saponification caused by cement dust. BS 5493
created resentment by its upgrading of metal coating

durability potential at the expense of paint. It is readily acknowledged that galvanizing and metal spraying provide a vital specific need but, considering economics and adaptability, painting must remain the general work-horse of the protection industry.

It is prudent to reflect that although good metal protection coated with chemical resistant paint such as chlorinated rubber or epoxy pitch may be superior to straight paint, the consequences of failure are disastrous in comparison. A major criticism of BS 5493 is its failure to highlight warnings illustrating what not to do and include case histories in confirmation.

MR G.K. WOOD, Messrs Sandberg

I think zinc spray is probably more convenient than aluminium in many cases, but it would be helpful in both methods for the sprayer to know what thickness he is applying. Perhaps a device could be introduced which gives a certain tone to the ear so long as the coating remains below the desired thickness.

In my experience the flame spray is the more reliable. Also if something breaks down, as it tends to do on a Sunday afternoon when the job is needed in a hurry, there is little one can do about a spray unless it is a gas spray. With the arc process it is probably necessary to call in a specialist and it may be days before one can start again.

DR R.D. JONES, University College, Cardiff

Hot dip galvanizing is the major process for continuously coated strip intended for subsequent fabrication. However, mention should be made of products like hot-dip aluminized steel strip and hot-dip terne (lead - 10% tin) coated steel strip. About a million tonnes of hot-dip aluminized steel is produced worldwide per annum; a significant amount of this is intended for the prefabricated building market. Aluminium coated steel is capable of giving 2-10 times the service of life of galvanized steel, the increase in lifetime depending on the exposure environment.[1]

What does Mr Porter think about the zinc-aluminium coated hot-dipped product known as Galvalume or Zincalume? The fact that Lysaght in Australia will have doubled their present annual zinc-aluminium coating capacity to 400 000 t by the end of 1979 indicates the importance the product has in some countries. Would hot dipped zinc-aluminium or aluminium coatings on steel provide the protection needed in the more severe Middle East environments?

MR I.P. GILLSON, Central Electricity Generating Board

The quality of work stems really from everyone knowing exactly what is required. This is laid down in the specification. I think it would be a great help if the civil engineering industry would accept practical specification clauses to which they could work, which could be passed on to operators and to inspectors.

A guarantee scheme tried by the CEGB was found unsatisfactory. It is difficult for penalty clauses to be implemented, mainly because if good paint is put on a proper surface in good conditions one would not expect a failure. The areas most likely to incur penalties on a guaranteed scheme seemed to be those specifically excluded by agreement between the employer and the contractor for good reasons.

It is a pity that the appointment of inspectors is often regarded as an extra cost to engineering or administration. There are really three costs: administration/engineering, the capital cost of the new work and the maintenance cost. No one worries if the capital cost tenders vary by 5-10%, but it is difficult for engineers to convey that a 5% cost of inspection will produce a benefit in excess of that in terms of extended coating life and reduced maintenance costs.

An increase in the interval of maintenance from 7 to 10 years would give a 42% return.

In certain industries the total cost of maintenance is increased by the costs of providing access and the penalties of taking economic installations out of service.

I suggest that to keep down the cost of inspection and yet make the best use of resources on an average civil engineering site, maximum use should be made of the clerks of works that are there. They can be the eyes and ears of the engineers and the technologists. I would like to see more short courses run and leaflets available to enable them to be more able to inspect the work as a member of the team.

On large jobs the appointment of a full-time specialist inspector is justified. I think it would be helpful to have a national qualification scheme for such inspectors to assist employers in making such appointments.

MR A.G. WARD, Babcock Corrosion Control Ltd

My company is involved in anti-corrosion protection, both as an applicator and as a fabricator. I think one must look at the commercial difficulties applicators face in bidding anti-corrosion finishes. For example, some years ago we acquired a business in the thin coatings field. We found that a significant contract had been sold at 60% of the fair price to a client in the public sector. The customer must have recognized that the price was unrealistic but was prepared to place an order on that basis.

Commercial pressures on applicators are therefore considerable. I would welcome specifications being applied more rigorously. My company would be prepared to make greater disclosure of the formulation of its elastomeric finishes so that there could be a better basis for comparison by the potential client.

There is a great difference between bidding to major contractors in the UK and to those overseas. In one case, my company was successful in gaining an American contract although our price was 15-20% above that of the competition. I do not believe the same thing would have happened in the UK. I believe that there are commerical pressures which militate against the applicator doing a thoroughly professional job.

MR C.A. PEQUIGNOT, Costain International Ltd

I take issue with the many criticisms that BS 5493 is of little use outside the UK. What do the critics expect? Section 5.1 of the code states 'the definitions ... and recommendations ... are primarily related to conditions in the UK'. Accepting that, it is, nevertheless, not difficult to make allowances for the fact that '... tropical environments can be much more corrosive than those in Britain'. At its simplest, the procedure can be described thus. Cut off the first column in each of parts 1-10 of table 3 and use it as a vertical slide-rule or nomogram of continuously variable parameters. Then, by relating such factors as the regional environment, local environment, miniclimates, length and conditions of sea transport, storage facilities and quality of site labour (all of which are familiar to an experienced overseas operator), the life column can be slid up or down accordingly as each factor is more or less unfavourable compared with UK conditions. (The process requires care and judgement but, given those, it will give a better result than working blindly to the code, adding something for luck and multiplying by a 'coefficient of so-whatness' (acknowledgement to Sir Harold Harding)). Careful reading of all relevant parts of the code and especially the notes to the tables enables elimination of incompatibilities and chemically unsuitable formulations. The specification arrived at thus is seldom much different from that of the experts.

Paint manufacturers are too fond of advertising a myriad types of paints under meaningless trade marks, then claiming to be the experts who must be asked, at an early stage,

DISCUSSION: PAPERS 9 AND 10

'Which does what?' Specifying might be easier, quicker, and
eventually cheaper, if labels on paint cans were to give
statements such as, 'Suitable for use as systems x, y, z (only)
in table 3 of BS 5493' or 'Formulated for use as product types
x, y, z in table 4 of BS 5493.' Such a labelling system would
give a more precise meaning to 'Caveat emptor'.

MR D. NORMAN, British Gas Corporation

The British Gas Corporation's painting inspection approval
scheme was set up in 1976 with the object of providing the
industry with a panel of approved inspectors. It is a three
tier system.

Any company can send an inspector to the Engineering
Research Station of British Gas and pay a fee to undergo a
written, oral and practical examination. On the overall
results of the examination the inspector is graded. A
vetting process is in continuous operation in the field. The
British Gas engineering staff report back on inspectors that
they use.

A scheme is now being set up by the Construction Industries
Training Board for national approval of painting inspectors.
It is hoped that it will become the norm for painting
inspectors to be certified and attend approved courses.

MR R.F. MANDER, Department of Transport

For some time I have been toying with the idea of impregna-
ting heavily corroded sections with resin when they are not
suitable for any other treatment. This can be done under
vacuum. I would only recommend this as a last resort and if
I were satisfied that there was enough sound metal left to
carry a load.

MR K.A. CHANDLER, British Steel Corporation

In research on the mechanism of rusting my firm has impreg-
nated rust with plastic in order to polish it and look at it
under the microscope. It is difficult to impregnate the rust
fully, but in certain cases if carefully done, although
expensive, it could be a method of treating it.

MR B.H. LEIGH-BRAMWELL, W. & J. Leigh & Co

Generally paint manufacturers are concerned when they see
specifications involving galvanized metal because the overcoating of galvanized metal is difficult. This is often
because different galvanizers have different processes, some
have different treatments of their galvanizing, and often

CORROSION IN CIVIL ENGINEERING

types of finished surface are completely different. Has
Mr Porter any solutions to these problems?
    What are the costs and the risks involved with galvanizing
compared with those for painting?

MR A.J. OWEN, Shell Composites Ltd

I think it is important that Cor-Ten should not be used under
water. I have only had contact with it once, on pontoons,
ten years old of 3/8 in. thick sheet. The sheet was a mass
of deep pits and in some cases it was penetrated.
    If specifiers are forming a specification they should
decide whether or not they really believe they are going to
get that specification, and having decided for the specification they must see it is adhered to even if it involves the
cost of an inspector. If this is not done half the specification will result, in which case it is better to start
with half the specification and to enforce that application.
    If this approach is not taken contractors trying to do a
good job will always be penalized and probably lose the work.

MR A. CHARLTON, British Railways

In paragraph 27 of his Paper, Mr Porter suggests using a zinc
primer on welded galvanized metal. Is zinc rich primer as
good as galvanizing, and if so, why?

MR PORTER

Mr Timmins comments on the increased corrosion of galvanized
steel in very acidic conditions. Environmental control
measures have now eliminated the worst of the acidic atmospheric conditions in the UK and consequently lives of about
five years for the galvanized coating on steel - which were
experienced on a few particularly aggressive acidic environments - are much less likely to be encountered in the future.
Thousands of test samples using a method developed by Shaw[2]
show that in 85% of the country structural steel with
610 g/m^2 minimum coating would last more than 25 years, and
in many rural parts the coating would last more than twice
as long, as is shown by many CEGB towers erected over 40 years
ago. Moreover, in alkaline conditions (such as those mentioned by Mr Timmins in relation to cement dust) galvanized
steel gives excellent corrosion resistance and for this
reason is used to protect steel parts embedded in concrete,
including steel used for reinforcement and prestressing of
concrete.

Mr Timmins also refers to failures with metal sprayed coatings but these were actually failures of the paint applied over metal spray. Essentially aluminium is unsuitable in alkaline conditions (unlike zinc) but is more tolerant of acidic conditions than zinc.

As regards the merits of arc spraying versus flame spraying, the metal spraying contractor will choose the process more suited to the job. Arc spraying is a more sophisticated technique than flame spraying, as implied by Mr Wood, but for major jobs arc spraying can often provide the optimum technical and economic performances. The problems outlined by Mr Timmins are not general.

A practical test instrument for determining adequate thickness of metal spraying was developed some years ago by Metallisation Ltd but I do not know if it is in commerical use.

Mr Péquignot has largely answered the comments relating to BS 5493 and the appraisal of the various types of coating. Metal spraying and galvanizing are sold under their technical names for the simple reason that users know their reliability from past experience.

Several speakers queried the economics of paint compared with metal coatings. For structural steelwork metal spraying plus sealer is usually more economic even on first cost than blasting plus three of four coats of paint, and galvanizing (which has a longer life to first maintenance than most competitive protective systems) is cheaper than the alternative systems on steelwork with average thickness up to 20 mm. However, users are sometimes faced with excessive mark-up by intermediaries who have less experience of using metal coating sub-contractors. Also the great variety in quality of paints and of painting practice is shown in quotations which have been obtained for painting of structures: the top quotation can be three times as much as the lowest - a variation six times as large as that normally found in galvanizing because galvanizing is a much more consistent and reproducible finish. Mr Ward's experience of a paint contract being sold for 60% of the fair price is typical.

The reliability of galvanizing compared with painting has been shown by van Eijnsbergen[3]. Paint has greater adaptability than galvanizing in that it is easy to slap on a coat of paint to a structure while it is on site to overcome earlier delays or defects. However, blasting and a four-coat painting system takes a minimum of a week to process through the shop in order to allow drying time between coats, but hot dip galvanizing of the same steelwork can often be accomplished in one shop within an hour or two.

CORROSION IN CIVIL ENGINEERING

Dr Jones' comments regarding aluminium coated steel and zinc-aluminium products are most relevant at present. In metal spraying, the anti-corrosion market is shared between aluminium spraying and zinc spraying; a 65% zinc-35% aluminium coating was used on a pilot scheme on sections of the Severn Bridge and elsewhere but production costs for wire, practical identification problems in store and uneconomic disposal of over-spray prevented this coating being developed, particularly as test work on the anti-corrosion properties of the coating both alone and when overcoated with paint did not show conclusive overall benefits compared with zinc or aluminium. Recently, the French and Belgians have developed the 85% zinc-15% aluminium alloy for spraying and this apparently has a useful market.

When considering the coatings on continuously processed steel strip the 55% aluminium-43½% zinc-1½% silicon alloy, known as Galvalume or Zincalume, has shown 1½-4 times the corrosion resistance of the thin galvanized coating on steel sheet in tests in Australia and the USA, but no long-term results are yet available from the very aggressive industrial atmospheres of Europe. Also there is some reason to believe that as little as 5% aluminium would give much of the same corrosion advantage provided appropriate processing techniques were available.

With organic materials on top of metal coatings a life of up to three times that expected from a simple sum of the separate lives can be obtained. The problem for associations like the Zinc Development Association, and indeed for many specifiers, is that whereas the metallic coatings can be uniquely specified by reference to published standards, most of the organic materials are sold under trade names and the contents of a trade named product can change overnight without warning. Moreover, it is very difficult to specify the contents of an organic material in such a way as to ensure that it will give the same performance all the time. This is particularly so with organic materials suitable for direct application to galvanized steel and in such cases the paint company and the galvanizer should have worked together on a combined system. Usually, however, the galvanized steel can be first treated with a phosphate type coating or an etch primer and then a wide range of paints can be applied. The Paintmakers Association and the Galvanizers Association[4] have agreed on the painting of galvanized steel; the practice recommended is relatively foolproof and allows for wide variation in the paints and in the surfaces which are being painted

In answer to Mr Charlton, the abrasion resistance of a zinc

rich primer is not nearly as good as galvanizing because the galvanized coating includes alloy layers which are very abra-sion resistant. For corrosion protection, particularly over small areas which might appear as a result of edges being cut, zinc dust paint is a good means of adding protection and in many cases extremely long lives can be obtained with quite thin coatings of zinc rich paint. The one which is the most widely quoted I think is a 30 year old pipeline in Australia. Modern zinc dust systems are now used in the UK which have been exposed in fairly aggressive conditions for 15-20 years and are behaving well. Certainly I regard zinc rich coatings as a viable means of protection in many cases where metal spraying or galvanizing is not suitable.

MR BAYLISS

I agree and sympathize with Mr Ward's comments on the unnecessary commercial pressures applied by the acceptance of the lowest tender without a realistic appraisal and comparison of the offers. In this connection it is vital that if it is intended to use full-time independent inspection on a coating operation, this fact must be known to the tenderers. They are then aware that they will not be able to cut corners.

I agree with Mr Péquignot's remarks in that a specification must say what it means and mean what it says. Such a specification makes the work of a quality control inspector much easier. Problems arise when the specification is imprecise and he has to make his own interpretation of limits.

In reply to Mr Norman, the Construction Industry Training Board is starting courses for painting inspectors. Each course takes one week and is intended for those that already have had field experience. There is a strong emphasis on the practical side, including knowledge of surface preparation and application techniques. This is because it is felt that an inspector should know more than his own inspection technique and should also be thoroughly familiar with the process he is monitoring. At the course centre there are excellent facilities for painting and blast cleaning.

It is hoped that an inspector approval scheme, possibly under the auspices of the Institution of Corrosion Science and Technology, will be run in conjunction with the courses.

References

1. Legault, R.A. and Pearson, V.P. The atmospheric corrosion of galvanised and aluminised steel. Inland Steel, East Chicago, 1976.

181

2. Shaw, T.R. Corrosion map of the British Isles. In Atmospheric factors affecting the corrosion of engineering metals. American Society for Testing and Materials, Philadelphia, 1978, ASTM STP 646, 204-215.

3. van Eijnsbergen, J.F.H. Reliability of hot dip galvanizing versus two paint systems. Address to Galvanizers Association General Meeting, 19 April, 1979.

4. Paintmakers Association and Galvanizers Association. General recommendations for painting galvanized steelwork. Paintmakers Association and Galvanizers Association, London, 1979.